MAGNETORHEOLOGICAL FLUID TECHNOLOGY

APPLICATIONS IN VEHICLE SYSTEMS

MAGNETORHEOLOGICAL FLUID TECHNOLOGY

APPLICATIONS IN VEHICLE SYSTEMS

SEUNG-BOK CHOI · YOUNG-MIN HAN

CRC Press
Taylor & Francis Group
Boca Raton   London   New York

CRC Press is an imprint of the
Taylor & Francis Group, an **informa** business

CRC Press
Taylor & Francis Group
6000 Broken Sound Parkway NW, Suite 300
Boca Raton, FL 33487-2742

Printed in the United States of America on acid-free paper
Version Date: 20120530

International Standard Book Number: 978-1-4398-5673-4 (Hardback)

Library of Congress Cataloging-in-Publication Data

Choi, Seung-Bok.
  Magnetorheological fluid technology : applications in vehicle systems / Seung-Bok Choi, Young-Min Han.
       p. cm.
  Summary: "This book represents recent research and developments on vehicle applications of smart materials using magnetorheological fluids (MR fluids), a type of smart fluid in a carrier fluid. It introduces physical phenomena and properties of MR fluids with potential applications, covers control methodologies, introduces the hysteresis identification of MR fluid and its application, and presents optimal design methods. It also covers applications for passenger and commercial vehicles, and includes case studies and examples related to the topics. For each application, a consecutive process is presented such as actuator design, system modeling, analysis, control strategy formulation, experimental set-up, and control performance"-- Provided by publisher.
  Includes bibliographical references and index.
  ISBN 978-1-4398-5673-4 (hardback)
  1. Motor vehicles--Electric equipment. 2. Electrorheological fluids. 3. Liquids--Magnetic properties. I. Han, Young-Min. II. Title.

TL272.C5227 2012
629.2--dc23
                                                                                           2012017948

Visit the Taylor & Francis Web site at
http://www.taylorandfrancis.com

and the CRC Press Web site at
http://www.crcpress.com

Contents

Preface

In recent years, smart materials technologies have been spreading rapidly and various engineering devices employing such technologies have been developed. The inherent characteristics of smart materials are actuator capability, sensor capability, and control capability. There are many smart material candidates that exhibit one or multifunctional capabilities. Among these, magnetorheological (MR) fluids, piezoelectric materials, and shape memory alloys have been effectively exploited in various engineering applications.

This book is a compilation of the authors' recent work on the application of MR fluids and other smart materials to use in vehicles. In particular, this book attempts to thread together the concepts that have been separately introduced through papers published by the authors in international, peer-reviewed journals. This book consists of nine chapters. In Chapter 1, we introduce the physical phenomenon and properties of MR fluids, and their potential applications. In Chapter 2, we discuss control methodologies that can be used to effectively control vehicle devices or systems featuring MR fluids. In Chapter 3, we introduce the hysteresis identification of MR fluid and its application through the adoption of the Preisach and polynomial models. In Chapter 4, we discuss an optimal design method and damping force control of MR shock absorber, which has practical applications in passenger cars. In addition, we introduce full-vehicle test results of a suspension system equipped with MR fluids. Chapter 5 discusses the application of MR-equipped suspension systems to tracked and railway vehicles. We evaluate their performance metrics (vibration controllability, position controllability, and stability) by using a controllable MR damper. Chapter 6 discusses potential application of MR technology to passenger vehicles. This chapter first introduces dynamic modeling and vibration control of an MR engine mount system associated with a full-car model, followed by a discussion of a novel MR impact damper positioned inside car bumpers to mitigate collision force. Chapter 7 discusses MR brake systems applicable to various classes of vehicles including passenger vehicles, motorcycles, and bicycles. This chapter deals with two types of brake mechanisms—bi-directional brakes for braking vehicles and torsional brakes for absorbing torsional vibrations. In Chapter 8, we discuss potential applications of MR technology for heavy vehicles. In this chapter, a drum-type MR fan clutch is introduced to actively control the temperature in engine rooms of commercial vehicles. Another application, a controllable MR seat damper, is introduced by presenting modeling and control strategies. In Chapter 9, we present two cases where haptic technologies are applied to vehicles. The first application is a multi-functional MR control knob for the easy operation of vehicle instruments such as the radio and air conditioning. The second application is a haptic cue system associated with accelerator pedals, which has been devised using MR fluids to achieve optimal gear shifting; we demonstrate experimentally its effectiveness and utility.

This book can be used as a reference text by graduate students who are interested in dynamic modeling and control methodology of vehicle devices, or systems associated with MR fluid technology. The students, of course, should have some technical and mathematical background in vibration, dynamics, and control in order to effectively master the contents. This book can also be used as a professional reference by scientists and engineers who wish to create new devices or systems for vehicles featuring controllable MR fluids.

The authors owe a debt of gratitude to many individuals; foremost is Professor N. M. Wereley at the University of Maryland who has collaborated with the authors in recent years in the field of smart materials. We acknowledge the contributions of many talented graduate and doctoral students at the Smart Structures and Systems Laboratory, Department of Mechanical Engineering, Inha University. Many of the experimental results presented in this book are the consequence of research endeavors funded by various agencies. In particular, the authors wish to acknowledge the financial support provided by the Korea Agency for Defense Development (Program Monitor Dr. M. S. Suh), the National Research Foundation of Korea (NRF), and Inha University's Research Fund.

Seung-Bok Choi and Young-Min Han

The Authors

Seung-Bok Choi received his PhD in mechanical engineering from Michigan State University, East Lansing in 1990. Since 1991, he has been a professor at Inha University, Incheon, South Korea. Currently, he is an Inha Fellow Professor, and his current research interests include the design and control of functional structures and systems utilizing smart materials such as electrorheological fluids, magnetorheological fluids, piezoelectric materials, and shape memory alloys. He is the author of over 310 archival international journals, 5 book contributions, and 220 international conference publications. He is currently serving as the associate editor of the *Journal of Intelligent Material Systems and Structures, Smart Materials and Structures,* and is a member of the editorial board of the *International Journal of Vehicle Autonomous Systems* and the *International Journal of Intelligent Systems Technologies and Applications.*

Young-Min Han received his PhD in mechanical engineering from Inha University, Incheon, South Korea in 2005. Since 2011, he has been a professor at Ajou Motor College, Boryeong, South Korea. His current research interest includes the design and control of functional mechanisms utilizing smart materials such as active mounts, semi-active shock absorbers, hydraulic valve systems, robotic manipulators, and haptic interfaces. Professor Han is the author of over 50 archival international journal articles and 25 international conference publications.

1

Magnetorheological Fluid

1.1 Physical Properties

The initial discovery and development of magnetorheological (MR) fluids is attributed to Jacob Rabinow at the U.S. National Bureau of Standards in the late 1940s [1–3]. Interestingly, even though MR fluids were introduced almost at the same time as electrorheological (ER) fluids, more patents and publications were reported in the late 1940s and early 1950s for MR fluids than for ER fluids [4]. Until recently, the non-availability of MR fluids of an acceptable quality has resulted in a dearth of relevant published literature, except for the brief flurry of publications in the period following their initial discovery. Encouragingly, there has been a resurgence of interest in MR fluids in recent years.

MR fluids belong to a family of rheological materials that undergo rheological phase-change under the application of magnetic fields. Typically, MR fluids are composed of soft ferromagnetic or paramagnetic particles (0.03~10 μm) dispersed in a carrier fluid. As long as the magnetizable particles exhibit low levels of magnetic coercivity, many different ceramic metal and alloys can be used in the composition of MR fluids. Usually, the MR particles are pure iron, carbonyl iron, or cobalt powder and the carrier fluid is a non-magnetic, organic, or aqueous liquid, usually a silicone or mineral oil. In the absence of a magnetic field, the MR particles are randomly distributed in the fluid. However, under the influence of an applied magnetic field, the MR particles acquire a dipole moment aligned with the external field and form chains, as shown in Figure 1.1. This chain formation induces a reversible yield stress in the fluid. In addition, the yield stress of the MR fluid is continuously and rapidly adjustable because it responds to the intensity of the applied magnetic field. As a result, MR fluid-based devices have inherent advantages such as continuously variable dynamic range and fast response.

From the fluid mechanics point of view, the behavior of MR fluid in the absence of a magnetic field can be described as Newtonian, while it exhibits distinct Bingham behavior in the presence of the field [5]. Therefore, MR

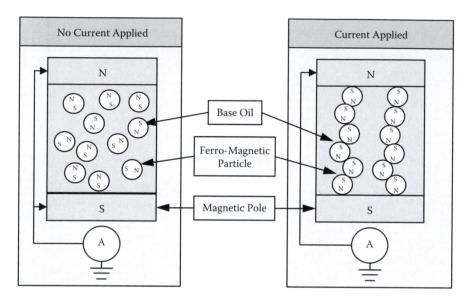

FIGURE 1.1
Microstructure of MR fluids.

fluid has been modeled in general as a Bingham fluid whose constitutive equation is given by the following:

$$\tau = \tau_y(\cdot) + \eta\dot{\gamma} \tag{1.1}$$

where η is the dynamic viscosity, γ is the shear rate, and $\tau_y(\cdot)$ is the dynamic yield stress of the MR fluid. It should be noted that the applied magnetic field could be expressed by either magnetic flux density (B) or magnetic field strength (H). Figure 1.2 presents the nature of the change from Newtonian to Bingham behavior. The dynamic yield shear stress (τ_y) increases as the magnetic field (H) increases. Under the magnetic potential, the total shear stress consists of two components—viscous-induced stress and field-dependent yield shear stress. The former is proportional to the shear rate, while the latter has an exponential relationship to the electric field. In order to quantitatively evaluate the field-dependent yield shear stress of MR fluid, two rheological regions of MR fluid are often adopted, as shown in Figure 1.3. MR fluid behaves like a linear viscoelastic material in the pre-yield region, a non-linear viscoelastic material in the yield region, and a plastic material in the post-yield region. The investigation of rheological properties in the yield and post-yield regions is very important in the design of MR application devices like dampers, mounts, valves, and clutches. The field-dependent dynamic yield shear stress (τ_y) is a significant property to be considered in such application devices. The field-dependent complex modulus (G^*) is a significant property in the

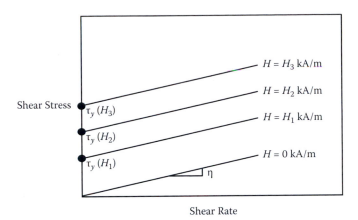

FIGURE 1.2
Bingham behavior of MR fluids.

pre-yield region, and it has been used to design smart structures for noise control and shock wave isolation. In general, the wave motion of the MR fluid domain in smart structures produces a small strain in the pre-yield region.

A rotational shear-mode type viscometer is generally used to measure yield shear stress and current density of the MR fluid. Figure 1.4 shows a parallel disk rheometer designed to measure the characteristics of the MR fluid.

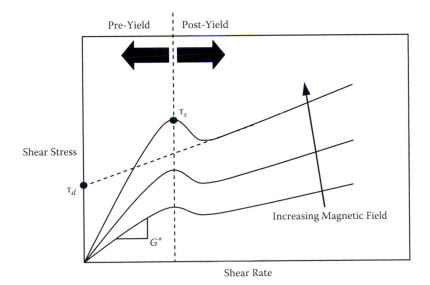

FIGURE 1.3
Shear stress–shear rate relationship of MR fluids.

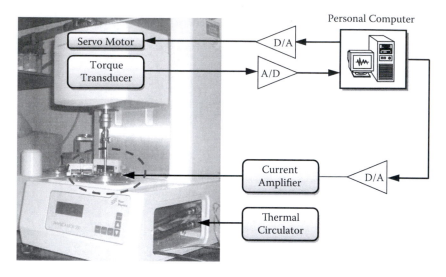

FIGURE 1.4
Experimental apparatus for Bingham characteristics measurement.

The MR cell has a magnetic core that produces a magnetic field. A gap exists between the rotating disk and the stationary plate, and this gap contains the MR fluid. In order to control the shear rate, a DC servomotor controls the rotational speed of the disk. A torque transducer is installed to measure the produced torque. The measured torque is then transmitted through an analog-to-digital (A/D) converter to a personal computer, which then calculates the corresponding shear stress. A rotary oscillation test with a dynamic spectrometer is necessary to determine the field-dependent viscoelastic properties of MR fluid with a complex shear modulus. Figure 1.5 shows a parallel-plate fixture designed for a rheometric mechanical spectrometer [6–13]. In this instrument, transducers attached to the bottom plate measure torque and normal force while the servo drives the top plate. MR fluid is then placed in the gap between the circular parallel plate fixtures and a magnetic field is applied to the gap. By sweeping oscillatory exciting frequencies of the fixtures, the complex shear modulus is measured under each constant magnetic field.

MR fluids are magnetic analogues of ER fluids and thus their essential rheological characteristics are similar to those of ER fluids. However, both MR and ER fluids possess a number of distinctive features. The achievable yield stress of MR fluids is an order of magnitude greater than that of ER fluids (2~10 kPa for ER fluids, 50~100 kPa for MR fluids), and MR fluids are not sensitive to impurities or contaminants that may be encountered during their manufacture and operation [14–16]. MR fluid-based devices have power requirements of below 50 Watts: 12 to 24 V and 1 to 2 A. Conventional batteries can readily supply this quantum of power.

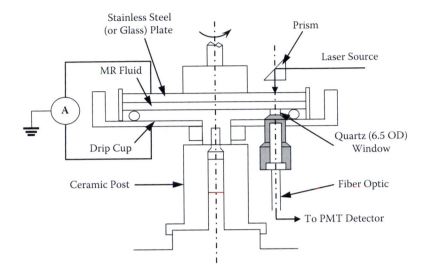

FIGURE 1.5
The parallel-plate fixture designed for a rheometric mechanical spectrometer.

The synthesis of high performance MR fluids requires that due consideration be given to several factors. First, advanced MR fluids must possess high yield stress at maximum magnetic fields and low viscosity in the absence of a field. Further, the MR effect should be stable within a wide temperature range. In addition to efforts to improve field-dependent yield stress, research on sedimentation, incompressibility, specific heat transfer property, wear aspect, fatigue property, and lubrication characteristics needs to be undertaken to aid the commercial development of advanced MR fluids. From the control perspective, the most important aspect to consider is the fluid's response time to the magnetic field. The fluid time constant needs to be quantitatively identified with respect to particle size, particle shape, viscosity of the base oil, conductivity of the particle, and so on. Determination of the fluid time constant is key to establishing a dynamic model of a fluid-based actuator, and in implementing a real-time feedback control scheme for MR devices. It has been reported that the response time of commercially available MR fluid is below 3 ms. This swift response can cover most dynamic systems in vehicles such as vehicle suspension and engine mount.

1.2 Potential Applications

When MR fluid is used as a controllable actuator in application devices, the working behavior of the fluid is classified into three different modes, as shown in Figure 1.6 [17, 18]. In the shear mode, it is normally assumed that

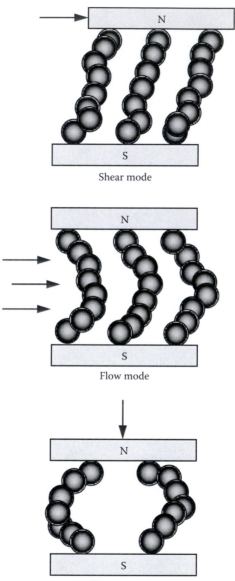

Shear mode

Flow mode

Squeeze mode

FIGURE 1.6
Operating modes of MR fluids.

one of the two field activation parts is free to translate or rotate about the other. Application devices operated in the shear mode include clutches [19–28], brakes [19, 29, 30], and vibration isolation mounts for small magnitude excitation [31–34]. Figure 1.7 depicts rotational type MR clutches that can be used for torque transmission or tension control in many dynamic systems.

FIGURE 1.7
Photograph of MR clutch.

The transmitted torque is easily controlled by adjusting the field intensity applied to the MR fluid domain. In the flow mode, it is assumed that the two field activation parts are fixed. There are essentially two control volumes in this mode, and the actuating force is controlled by adjusting the pressure difference between the two control volumes. Application devices operated in the flow mode include shock absorbers for vehicle suspension systems [35–61], seat dampers [62–66], recoil dampers [67–69], landing gear [70,71], large stroke dampers for bridges and buildings [72–74], seismic dampers [75,76], and vibration isolation mounts for large magnitude excitation [77]. Figure 1.8 depicts an MR damper that can be used in passenger vehicles. The pressure difference between the two chambers of the damper is controlled by adjusting the field intensity under various road conditions. Thus, the desired damping force for the suspension system can be obtained under rough road conditions, thereby providing both passenger comfort and driving stability. Yet another application of the flow mode type damper is in medical devices, such as the prosthesis shown in Figure 1.9. Most MR dampers operated in the flow mode are semi-active, wherein only energy dissipation increases through the application of a field to the fluid domain. However, by constructing hydraulic valve systems that also operate in the flow mode, the MR fluid can be used as an active actuator [78–81]. In the MR hydraulic valve system depicted in Figure 1.10, the pressure or flow rate can be controlled by adjusting the field intensity. By doing this, one can effectively control the position and velocity of the cylinder systems that are normally integrated with such hydraulic systems. MR hydraulic valve systems can replace conventional servo-valve systems in various application fields [82, 83]. Unlike in the two

FIGURE 1.8
Photograph of MR shock absorber.

operating modes previously introduced, in the squeeze mode the activation gap is varied in the vertical direction. In this mode, the MR fluid is squeezed by a normal force; hence, both tensile and compressive forces can be controlled by the field intensity. Application devices operated in the squeeze mode include vibration isolation systems for small magnitude excitations, such as the mount [84] depicted in Figure. 1.11. MR fluids can be also embedded into conventional host structures to make the so-called smart structures

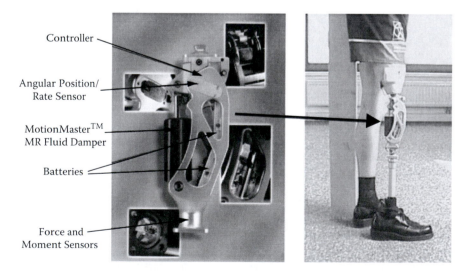

FIGURE 1.9
Photograph of MR damper for a leg.

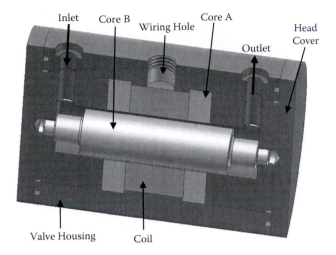

FIGURE 1.10
Photograph of the MR valve.

shown in Figure 1.12. By doing this, the complex modulus of the smart structure can be controlled by adjusting the field intensity. This, in turn, helps control the modal characteristics of the smart structure such as natural frequencies and mode shapes. Thus, unwanted vibrations or noises occurring due to resonance or external excitations can be effectively controlled by synthesizing appropriate smart structure systems. Smart structure applications associated with MR fluids include dash panels, aircraft wings, helicopters, and flexible robotic arms [85–91].

FIGURE 1.11
Photograph of the MR mount.

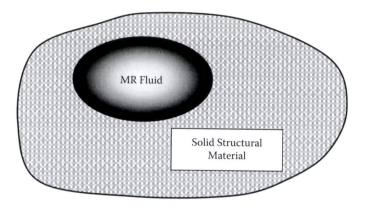

FIGURE 1.12
Generic idea of smart structures featuring MR fluid.

Most application devices employing MR fluid technology need to be integrated with a semi-active or active control scheme. In order to produce commercially viable MR devices, reliable design methodology, accurate dynamic modeling, and efficacious control algorithms are of the essence. Reliable design methodology implies simple design, compact size, and high durability, as well as easy and affordable operation. Simplicity in design can be achieved by reducing the number of components and eliminating moving parts, if possible. The size of the mechanism is very important for practical implementation in real environments. Optimization can be achieved by considering the MR effect with respect to design parameters such as field activated gap and length. The durability of an MR mechanism has two separate aspects—mechanical durability and electrical or magnetic durability. Mechanical durability consists of measuring fatigue resistance, anti-wear, and leakage resistance of the components during the long-cycle of motion. Magnetic durability indicates how long the MR mechanism can remain operational against the employed magnetic field without any breakdown. This may depend on the uniformity of the gap size during dynamic motion, and on the materials used. Ease of use is essential not only for the basic operating mechanism, but also for the maintenance and replacement of faulty components and the MR fluid refill process. These design goals can be achieved by accurately analyzing the constitutive behaviors of the fluid. In other words, accurate analysis and dynamic modeling for MR devices are necessary to arrive at reliable design specifications.

In general, MR fluid-based control systems consist of a set of sensors, signal converters, a microprocessor, a voltage isolator, a current amplifier, and a control algorithm. Most currently available sensors, such as accelerometers, can be adapted to measure the dynamic response of the devices. The microprocessor, which includes A/D and D/A (digital-to-analog) signal converters, plays a very important role in the closed-loop control time. The microprocessor should have at least 12 bits to store and operate the control software, and should be able to accommodate high sampling frequencies up to 10 kHz. Furthermore, the response time of the current amplifier to the source input voltage should be fast enough to facilitate the working of the feedback control system. Most control algorithms currently used to attenuate vibration using MR fluids are dubbed semi-active. The performance of semi-active control systems can generally be enhanced in the active mode without the use of large power sources. One of the most popular control logics for semi-active control systems is the skyhook control algorithm because this is easy to formulate and implement in practice. Groundhook controllers, sky–groundhook controllers, and neuro-fuzzy skyhook controllers are also semi-active control schemes that can be applied to MR control devices. Possible active controller candidates for semi-active control systems are the PID (proportional–integral–derivative) controller, the LQG (linear quadratic Gaussian) controller, the sliding mode controller, the H-infinity controller, and the Lyapunov-based state feedback controller. Since semi-active actuators cannot increase the mechanical energy of the control system, special attention must be paid while adapting active control strategies to semi-active control systems. On the other hand, an active control system can be constructed by synthesizing a hydraulic MR valve-cylinder system in a closed-loop manner. In this case, currently available active control strategies can be applied to the MR control system without any modification.

References

[1] Rabinow, J. 1948. The magnetic fluid clutch. *AIEE Transactions* 67: 1308-1315.
[2] Rabinow, J. 1948. Magnetic fluid clutch. *National Bureau of Standards Technical News Bulletin* 32: 54–60.
[3] Rabinow, J. 1951. Magnetic fluid torque and force transmitting device. United States Patent 2.
[4] Jolly, M. R., Jonathan, W. B., and Carlson, J. D. 1999. Properties and applications of commercial magnetorheological fluids. *Journal of Intelligent Material Systems and Structures* 10: 5–13.
[5] Carlson, J. D., Catanzarite, D. M., and St. Clair, K. A. 1996. Commercial magneto-rheological devices. *International Journal of Modern Physics B* 10: 2857–2865.
[6] Li, W. H., Chen, G., and Yeo, S. H. 1999. Viscoelastic properties of MR fluids. *Smart Materials and Structures* 8: 460–468.
[7] Claracq, J., Sarrazin, J., and Montfort, J. P. 2003. Viscoelastic properties of magnetorheological fluids. *Rheologica ACTA* 43: 38–49.
[8] Wang, Y. 2006. Modeling of polymer melt/nanoparticle composites and magneto-rheological fluids, Ph. D. Dissertation, Ohio State University, Columbus.

[9] Wei, B., Cong, X., and Jiang, W. 2010. Influence of polyurethane properties on mechanical performances of magnetorheological elastomers. *Journal of Applied Polymer Science* 116: 771–778.

[10] Waigh, T. A. 2005. Microrheology of complex fluids. *Reports on Progress in Physics* 68: 685–743.

[11] Wang, X. and Gordaninejad, F. 2007. Flow analysis and modeling of field-controllable, electro-and magneto-rheological fluid dampers. *Journal of Applied Mechanics* 74: 13–22.

[12] Li, W. H., Chen, G., Yeo, S. H., and Du, H. 2002. Dynamic properties of magnetorheological materials. *Key Engineering Materials* 227: 119–124.

[13] Or, S. W., Duan, Y. F., Ni, Y. Q., Chen, Z. H., and Lam, K. H. 2008. Development of magnetorheological dampers with embedded piezoelectric force sensors for structural vibration control. *Journal of Intelligent Material Systems and Structures* 19: 1327–1338.

[14] Carlson, J. D. and Weiss, K. D. 1994. A growing attraction to magnetic fluids. *Machine Design*, 61–64.

[15] Dyke, S. J., Spencer Jr., B. F., Sain, M. K., and Carlson, J. D. 1996. Experimental Verification of Semi-active Structural Control Strategies Using Acceleration Feedback. *Third International Conference on Motion and Vibration Control*, Chiba, pp. 1–6.

[16] Yang, G., Spencer Jr., B. F., Carlson, J. D., and Sain, M. K. 2002. Large-scale MR fluid damper: modeling and dynamic performance considerations. *Engineering Structure*. 24: 309–323.

[17] Boelter, R. and Janocha. H. 1997. Design rules for MR fluid actuators in different working modes. *Proceedings of the SPIEs 1997 Symposium on Smart Structures and Materials* 3045: 148–159.

[18] Wereley, N. M., Cho, J. U., Choi, Y. T., and Choi, S. B. 2008. IJ-209 Magnetorheological dampers in shear mode. *Smart Materials and Structures* 17: 1–11.

[19] Hongsheng, H., Hongsheng, H., Juan, W., Liang, C., Jiong, W., and Xuezheng J. 2009. *Proceedings of the IEEE International Conference on Automation and Logistics*, Shenyang, China, pp. 1248–1253.

[20] Choi, S. B., Hong, S. R., Cheong, C. C., and Park, Y. K. 1999. Comparison of field-controlled characteristics between ER and MR clutches. *Journal of Intelligent Material Systems and Structures* 10: 615–619.

[21] Saito, T. and Ikeda, H. 2007. Development of normally closed type of magnetorheological clutch and its application to safe torque control system of human-collaborative robot. *Journal of Intelligent Materials Systems and Structures* 18: 1181–1185.

[22] Jin, H., Lin-qing, L., and Chang-hua, L. 2006. Design analysis of cylindrical magnetorheological clutch. *Journal of Functional Materials* 5: 760–761.

[23] Lee, U., Kim, D., Hur, N. G., and Jeon, D. Y. 1999. Design analysis and experimental evaluation of an MR fluid clutch. *Journal of Intelligent Material Systems and Structures* 10: 701–707.

[24] Kavlicoglu, B. M., Gordanieljad, F., Evrensel, C. A., Liu, Y. Kavlicoglu, N., and Fuchs, A. 2008. Heating of a high-torque magnetorheological fluid limited slip differential clutch. *Journal of Intelligent Material Systems and Structures* 19: 235–241.

[25] Neelakantan, V. A. and Washington, G. N. 2005. Modeling and reduction of centrifuging in magnetorheological (MR) transmission clutches for automotive applications. *Journal of Intelligent Material Systems and Structures* 16: 703–711.

[26] Jedryczka, C., Piotr, S., and Wojchiech, S. 2009. The influence of magnetic hysteresis on magnetorheological fluid clutch operation. *International Journal for Computation and Mathematics in Electrical and Electronic Engineering* 28: 711–721.

[27] Kavlicoglu, N. C., Kablicoglu, B. M., Liu, Y., Evrensel C. A., Fuchs, A. Korol, G., and Gordaninejad F. 2007. Response time and performance of a high-torque magneto-rheological fluid limited slip differential clutch. *Journal of Smart Material Systems and Structures* 16: 149–159.

[28] Hakogi, H., Ohaba, M., Kuramochi, N., and Yano, H. 2005. Torque control of a rehabilitation teaching robot using magneto-rheological fluid clutches. *JSME International Journal series B.* 48: 501–507.

[29] Li, W. H. and Du, H. 2003. Design and experimental evaluation of a magnetorheological brake. *International Journal of Advanced Manufacturing Technology* 21: 508–515.

[30] Karakoc, K., Park, E. J., and Suleman, A. 2008. Design considerations for an automotive magnetorheological brake. *Mechatronics* 18: 434–447.

[31] Hong, S. R. and Choi, S. B. 2005. Vibration control of a structural system using magneto-rheological fluid mount. *Journal of Smart Material Systems and Structures* 16: 931–936.

[32] Choi, S. B., Hong, S. R., Sung, K. G., and Sohn, H. W. 2008. Optimal control of structural vibrations using a mixed-mode magnetorheological fluid mount. *International Journal of Mechanical Sciences* 50: 559–568.

[33] Ahmadian, M. and Ahn, Y. K. 1999. Performance analysis of magneto-rheological mounts. *Journal of Smart Material Systems and Structures* 10: 248–256.

[34] Ahn, Y. K., Ahmadian, M., and Morushita, S. 1999. On the design and development of a magneto-rheological mount. *Vehicle System Dynamics: International Journal of Vehicle Mechanics and Mobility* 32: 199–216.

[35] Yao, G. Z., Yap, F. F., Chen, G., Li, W. H., and Yeo, S. H. 2002. MR damper and its application for semi-active control of vehicle suspension system. *Mechatronics* 12: 963–973.

[36] Yu, M., Dong, X. M., Choi, S. B., and Liao, C. R. 2009. Human simulated intelligent control of vehicle suspension system with MR dampers. *Journal of Sound and Vibration* 319: 753–767.

[37] Du, H., Szeb, K. Y., and Lam, J. 2005. Semi-active H∞ control of vehicle suspension with magneto-rheological dampers. *Journal of Sound and Vibration* 283: 981–996.

[38] Choi, S. B., Lee, H. S., and Park, Y. P. 2002. H∞ control performance of a full-vehicle suspension featuring magnetorheological dampers. *Vehicle System Dynamics: International Journal of Vehicle Mechanics and Mobility* 38: 341–360.

[39] Lai, C. Y. and Liao, W. H. 2002. Vibration control of a suspension system via a magnetorheological fluid damper. *Journal of Vibration and Control* 8: 527–547.

[40] Yokoyama, M., Hedrick, J. K., and Toyama, S. 2001. A model following sliding mode controller for semi-active suspension systems with MR dampers. *Proceedings of the American Control Conference*, Arlington, VA, 4: 2652–2657.

[41] Choi, S. B. and Han, Y. M. 2003. MR seat suspension for vibration control of a commercial vehicle. *International Journal of Vehicle Design* 31: 202–215.

[42] Yu, M., Liao, C. R., Chen, W. M., and Huang, S. L. 2006. Study on MR semi-active suspension system and its road testing. *Journal of Intelligent Material Systems and Structures* 17: 801–806.

[43] Guo, D. L., Hu, H. Y., and Yi, J. Q. 2004. Neural network control for a semi-active vehicle suspension with a magnetorheological damper. *Journal of Vibration and Control* 10: 461–471.

[44] Lee, H. S. and Choi, S. B. 2000. Control and response characteristics of a magneto-rheological fluid damper for passenger vehicles. *Journal of Intelligent Material Systems and Structures* 11: 80–87.

[45] McManus, S. J., St. Clair, K. A., Boileau, P. É., Boutin, J., and Rakheja, S. 2002. Evaluation of vibration and shock attenuation performance of a suspension seat with a semi-active magnetorheological fluid damper. *Journal of Sound and Vibration* 253: 313–327.

[46] Liu, Y., Gordaninejad, F., Evrensel, C., Karakas, E. S., and Dogruer, U. 2004. Experimental study on fuzzy skyhook control of a vehicle suspension system using a magnetorheological fluid damper. *Proceedings of SPIE* 5388, 338.

[47] Wereley, N. M. and Pang, L. 1998. Nondimensional analysis of semi-active electrorheological and magnetorheological dampers using approximate parallel plate models. *Smart Materials and Structures* 7: 732–743.

[48] Liao, W. H. and Wang, D. H. 2003. Semiactive vibration control of train suspension systems via magnetorheological dampers. *Journal of Intelligent Material Systems and Structures* 14: 161–172.

[49] Lam, A. H. F. and Liao, W. H. 2003. Semi-active control of automotive suspension systems with magneto-rheological dampers. *International Journal of Vehicle Design* 33: 50–75.

[50] Song, X., Ahmadian, M., Southward, S., and Miller, L. R. 2005. An adaptive semi-active control algorithm for magnetorheological suspension systems. *Journal of Vibration and Acoustics* 127: 493–502.

[51] Batterbee, D. C. and Sims, N. D. 2007. Hardware-in-the-loop simulation of magnetorheological dampers for vehicle suspension systems. *Journal of Systems and Control Engineering* 221: 265–278.

[52] Shen, Y., Golnaraghi, M. F., and Heppler, G. R. 2006. Semi-active vibration control schemes for suspension systems using magnetorheological dampers. *Journal of Vibration and Control* 12: 3–24.

[53] Choi, S. B., Lee S. K., and Park, Y. P. 2001. A hysteresis model for field-dependent damping force of a magnetorheological damper. *Journal of Sound and Vibration* 245: 375–383.

[54] Sassi, S., Cherif, K., Mezghani, L., Thomas, M., and Kotrane, A. 2005. An innovative magnetorheological damper for automotive suspension: from design to experimental characterization. *Smart Materials and Structures* 14: 811–822.

[55] Prabakar, R. S., Sujatha, C., and Narayanan, S. 2009. Optimal semi-active preview control response of a half car vehicle model with magnetorheological damper. *Journal of Sound and Vibration* 326: 400–420.

[56] Choi, S. B., Lee, H. S., and Park, Y. P. 2002. H-infinity control performance of a full-vehicle suspension featuring magnetorheological dampers. *Vehicle System Dynamics* 38: 341–360.

[57] Zhang, C. W., Ou, J. P., and Zhang, J. Q. 2006. Parameter optimization and analysis of a vehicle suspension system controlled by magnetorheological fluid dampers. *Structural Control and Health Monitoring* 13: 885–896.

[58] Guo, S., Li, S., and Yang, S. 2006. Semi-active vehicle suspension systems with magnetorheological dampers. *IEEE International Conference on Vehicular Electronics and Safety* (ICVES 2006), 403–406.

[59] Cao, J., Liu, H., Li, P., and Brown, D. J. 2008. State of the art in vehicle active suspension adaptive control systems based on intelligent methodologies. *IEEE Transactions on Intelligent Transportation Systems* 9: 392–405.

[60] Ahmadian, D. S. M. 2001. Vehicle evaluation of the performance of magneto rheological dampers for heavy truck suspensions. *Journal of Vibration and Acoustics* 123: 365–375.

[61] Dogruer, U., Gordaninejad, F., and Evrensel, C. A. 2008. A new magneto-rheological fluid damper for high-mobility multi-purpose wheeled vehicle (HMMWV). *Journal of Intelligent Material Systems and Structures* 19: 641–650.

[62] McManus, S. J. and St. Clair, K. A. 2002. Evaluation of vibration and shock attenuation performance of a suspension seat with a semi-active magnetorheological fluid damper. *Journal of Sound and Vibration* 253: 313–327.

[63] Choi, S. B., Nam, M. H., and Lee, B. K. 2000. Vibration control of a MR seat damper for commercial vehicles. *Journal of Intelligent Material Systems and Structures* 11: 936–944.

[64] Han, Y. M., Nam, M. H., Han, S. S., Lee, H. G., and Choi, S. B. 2002. Vibration control evaluation of a commercial vehicle featuring MR seat damper. *Journal of Smart Material Systems and Structures* 13: 575–579.

[65] Lee, Y. R. and Jeon, D. Y. 2002. A study on the vibration attenuation of a driver seat using an MR fluid damper. *Journal of Smart Material Systems and Structures* 13: 437–441.

[66] Choi, S. B. and Han, Y. M. 2003. MR seat suspension for vibration control of a commercial vehicle. *International Journal of Vehicle Design* 31: 201–215.

[67] Ahmadian, M. and Poynor, J. C. 2001. An evaluation of magneto rheological dampers for controlling gun recoil dynamics. *Shock and Vibration* 8: 147–155.

[68] Ahmadian, M., Appleton, R., and Norris, J. A. 2002. An analytical study of fire out of battery using magneto rheological damper. *Shock and Vibration* 9: 129–142.

[69] Facey, W., Rosenfeld, N., Choi, Y. T., Wereley, N., Choi, S. B., and Chen, P.C. 2005. Design and testing of a compact magnetorheological impact damper for high impulsive loads. *International Journal of Modern Physics Part B* 19: 1549–1555.

[70] Batterbee, D. C., Sims, N. D., Stanway, R., and Wolejsza, Z. 2007. Magnetorheological landing gear: A design methodology. *Smart Materials and Structures* 16: 2429–2440.

[71] Choi, Y. T. and Wereley, N. M. 2003. Vibration control of a landing gear system featuring electrorheological/magnetorheological fluids. *AIAA Journal of Aircraft* 40: 423–439.

[72] Spencer Jr., B. F., Dyke, S. J., Sain, M. K., and Carlson, J. D. 1997. Phenomenological model of a magnetorheological damper. *ASCE Journal of Engineering Mechanics* 123: 230–238.

[73] Yang, G. 2001. Large-scale magnetorheological fluid damper for vibration mitigation: modeling, testing and control. Ph.D. Thesis, University of Notre Dame.

[74] Liu, Y., Gordaninejad, F., Evrensel, C. A., Wang, X., and Hitchhock, G. 2005. Comparative study on vibration control of a scaled bridge using fail-safe magneto-rheological fluid dampers. *ASCE Journal of Structural Engineering* 131: 743–751.

[75] Hiemenz, G., Choi, Y. T., and Wereley, N. M. 2003. Seismic control of civil structures utilizing semi-active MR braces. *Journal of Computer-Aided Civil and Infrastructures Engineering* 18: 31–44.

[76] Dyke, S. J., Spencer Jr., B. F., Sain, M. K., and Carlson, J. D. 1996. Modeling and control of magnetorheological dampers for seismic response reduction. *Smart Materials and Structures* 5: 565–575.

[77] Choi, Y. T. and Wereley, N. M. 2005. Semi-active vibration isolation using magnetorheological isolators. *AIAA Journal of Aircraft* 42: 1244–1251.

[78] An, J. and Kwon, D. S. 1997. Modeling of a magnetorheological fluid (MRF) actuator using a magnetic hysteresis-based model. *Proc. of 14th Korean Automatic and Control Conference*, Yongin, South Korea, pp. 219–222.

[79] Kim, K. H., Nam, Y. J., Yamane, R., and Park, M. K. 2009. Smart Mouse: 5-DOF haptic hand master using magneto-rheological fluid actuators. *Journal of Physics: Conference Series* 149: 1–6.

[80] Yoo, J. H., Siroshi, J., and Wereley N. M. 2005. A magnetorheological piezo-hydraulic actuator. *Journal of Intelligent Material Systems and Structures* 16: 945–953.

[81] Yoo, J. H. and Wereley, N. M. 2004. Performance of a magnetorheological hydraulic power actuation system. *Journal of Intelligent Material Systems and Structures* 15: 847–858.

[82] Nosse, D. T. and Dapino, M. J. 2007. Magnetorheological valve for hybrid electrohydrostatic actuation. *Journal of Intelligent Material Systems and Structures* 18: 1121–1136.

[83] Nguyen, Q. H., Choi, S. B., and Welerey, N. M. 2008. Optimal design of magnetorheological valves via a finite element method considering control energy and a time constant. *Smart Materials and Structures* 17: 1–12.

[84] Choi, S. B., Song, H. J., Lee, H. H., Lim, S. C., and Kim, J. H. 2003. Vibration control of a passenger vehicle featuring magnetorheological engine mount. *International Journal of Vehicle Design* 33: 2–16.

[85] Schweiger, J. 2002. Aircraft Control, Applications of Smart Materials, *Encyclopedia of Smart Materials*, Wiley Online Library.

[86] Chen, L. and Hansen, C. H. 2005. Active vibration control of a magnetorheological sandwich beam. *Proceedings of ACOUSTICS*, Busselton, Western Australia, pp.93–98.

[87] Marathe, S., Gandhi, F., and Wang, K. W. 1998. Helicopter blade response and aeromechanical stability with a magnetorheological fluid based lag damper. *Journal of Intelligent Material Systems and Structures* 9: 241–282.

[88] Wereley, N. M., Kamath, G. M., and Madhavan, V. 1999. Hysteresis modeling of semi-active magnetorheological helicopter lag damper. *Journal of Intelligent Material Systems and Structures* 10: 624–633.

[89] Zhao, Y., Choi, Y. T., and Wereley, N. M. 2004. Semi-active damping of ground resonance in helicopters using magnetorheological dampers. *Journal of the American Helicopter Society* 4: 468–492.

[90] Yoon, S. S., Kang, S. K., Yun, S. K., Kim, S. J., Kim, Y. H., and Kim, M. S. 2005. Safe arm design with MR-based passive compliant joints and visco-elastic covering for service robot applications. *Journal of Mechanical Science and Technology* 19: 1835–1845.

[91] Pettersson, A., Davis, S., Gray, J. O., Dodd, T. J., and Ohlsson, T. 2010. Design of a magnetorheological robot gripper for handling of delicate food products with varying shapes. *Journal of Food Engineering* 98: 332–338.

2

Control Strategies

2.1 Introduction

An automatic controller, in general, compares actual output of a plant with that of the reference input (desired value) and produces a control signal that will reduce the deviation between the actual and desired values to zero or a small value. Figure 2.1 presents a block diagram of a typical control system that is integrated with magnetorheological (MR) devices. The controller detects the actuating error signal, which is usually a very low power level, and amplifies it to a sufficiently high level via the current amplifier for MR devices. The actuator is a power device that produces plant input according to the control signal.

Conventional actuators, such as electric motors, are normally used in the active control system. However, the actuator based on MR fluid is usually employed in the semi-active control system. Using the active actuator type, control energy can be both taken away from and inserted into the plant, whereas with the semi-active actuator type, it can only be dissipated. MR devices can be used as active actuators via a hydraulic servovalve mechanism associated with the pump. Most currently available sensors, such as accelerometers, can be adapted to measure the dynamic control system responses associated with MR devices. In this chapter, some very effective control methodologies for vehicle application systems featuring MR devices are discussed.

2.2 Semi-Active Control

In general, control energy can be either taken away or inserted into the system to be controlled by employing an active control device. However, with a semi-active device, the control energy can be only dissipated [1]. So, this increases the degree of stability of the control system. Using MR devices, the damping of the system can be continuously controlled by controlling the

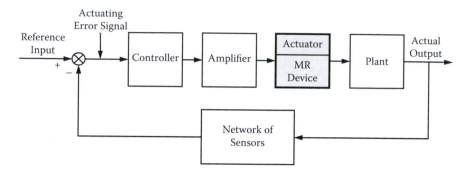

FIGURE 2.1
Block diagram of a typical control system featuring MR devices.

intensity of the applied field as shown in Figure 2.2. It is seen that desired damping force can be achieved in the controllable domain regardless of the velocity. This is a salient feature of the semi-active control system activated by MR devices. In order to achieve the desired damping force in the controllable domain, one can use three different semi-active control methods: skyhook, groundhook, and sky–groundhook controllers.

The skyhook controller is introduced by Karnopp et al. [2]. It is well known that the logic of the skyhook controller is simple and easy to implement to a real field. Figure 2.3 presents a schematic configuration of the skyhook controller. The desired damping force, F_d, can be set by

$$F_d = C_{sky}(E)\dot{x} \tag{2.1}$$

FIGURE 2.2
Semi-active control characteristics with MR devices.

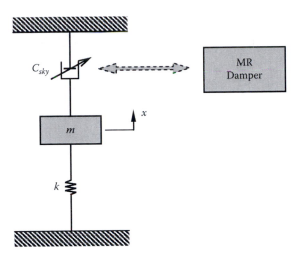

FIGURE 2.3
A semi-active system with the skyhook controller.

where C_{sky} is a control gain that physically indicates the damping. The control gain can be judiciously tuned by the magnetic field to meet the desired damping force. In practice, the control gain can be chosen by on-off logic as:

$$C_{sky} \text{ is maximum, } \dot{x} > 0$$

$$C_{sky} \text{ is minimum, } \dot{x} \le 0$$

(2.2)

The above on-off control logic of the skyhook controller is frequently adopted in vehicle suspension. The shock absorber in vehicle suspension should produce relatively high damping force in the rebound motion, while producing small damping force in the jounce motion. In other words, we can control hard mode (high damping) and soft mode (small damping) by just controlling the magnetic field without using the directional check valve. The skyhook control method is normally used to isolate the vibration of sprung mass, which is directly connected to ground. One can use a groundhook controller to isolate the vibration of unsprung mass, which is directly connected to ground. For example, the skyhook control is used to improve ride quality by suppressing the body (spring mass) motion, while the groundhook control is used to improve steering stability by suppressing the tire (unsprung mass) motion. The desired damping can be set the same as Equation (2.1).

If we want to improve both ride comfort and steering stability of the vehicle system, the sky–groundhook controller shown in Figure 2.4 can be used. The sky–groundhook controller consists of two ideal dampers; one is fixed to the ceiling, while the other is fixed to the ground. The one fixed to the ceiling produces a damping force to control the vibration of the vehicle body (sprung mass, m_s), while the one fixed to the ground generates a damping

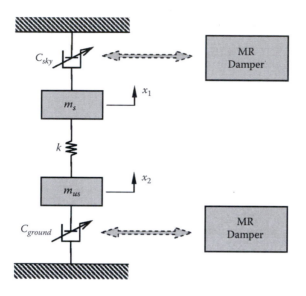

FIGURE 2.4
A semi-active system with the sky–groundhook controller.

force to control the vibration of the wheel (unsprung mass, m_{us}). Therefore, we may improve both the ride quality and the steering stability by properly adjusting each component of the damping forces associated with MR dampers. The desired damping force can be set by

$$F_d = \sigma C_{sky} \dot{x}_1 + (1 - \sigma)C_{ground} \dot{x}_2 \qquad (2.3)$$

where C_{sky} is the skyhook control gain, C_{ground} is the groundhook control gain, and σ $(0 \leq \sigma \leq 1)$ is the weighting parameter between two control inputs. It is noted that σ plays a crucial role for control performance. For instance, if σ is equal to zero, the damping force (control input) takes account for only the steering stability. Consequently, appropriate determination of the weighting parameter depending upon the dynamic motion of the vibrating structures or systems is to be absolutely required. One way to achieve this is to use a fuzzy algorithm [3, 4].

As an application example of the skyhook controller, consider the damping force control of an MR shock absorber shown in Figure 2.5. The damping force of the shock absorber can be derived by

$$F = k_g \dot{x}_p + c_e \dot{x}_p + F_{MR}(H) + F_f \qquad (2.4)$$

where k_g is the stiffness of the gas chamber, c_e is the equivalent viscous damping coefficient, F_{MR} is the controllable damping force, F_f is the frictional force, and $\dot{x}_p$ is the piston velocity. The controllable damping force is a function of the magnetic field H. Thus, by replacing actual damping force F by the desired damping force F_d, the required control magnetic field to obtain the

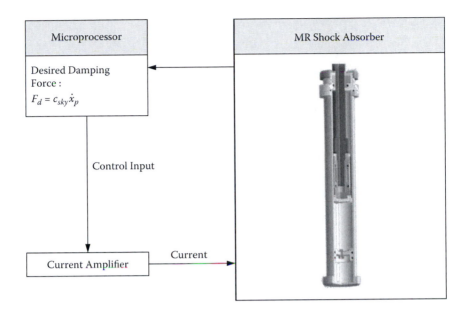

FIGURE 2.5
Skyhook damping force control of an MR shock absorber.

desired value is easily obtained. In general, for the practical implementation of the MR shock absorber to the vehicle suspension system we require high damping force in the rebound (extension) motion, while small damping force is required in the jounce (compression) motion.

In order to obtain this performance target, a conventional passive or semi-active damper has a directional check valve. This valve allows fluid flow between the upper and lower chambers only in the jounce motion, and hence decreases the damping force. However, in the MR shock absorber, we can achieve this performance goal with completely different means. In other words, we can control hard (high damping) and soft (small damping) mode by just controlling the magnetic field without using the directional check valve. In order to demonstrate this controllability, the following desired damping force is chosen:

$$\text{If } \dot{x}_p \geq 0 \text{ (rebound)}, \quad F_d = C\dot{x}_p, \ C = \text{control gain}$$

$$\text{If } \dot{x}_p < 0 \text{ (jounce)}, \quad F_d = F \text{ at zero field}$$

(2.5)

The above desired target describes that the MR shock absorber should produce relatively high damping force in the rebound motion, while producing small damping force in the jounce motion. The controlled (actual) responses for the above desired damping trajectories are presented in Figure 2.6 [5]. We clearly observe from the tracking response that favorable damping force tracking is achieved by controlling the magnetic field. It is evident from the

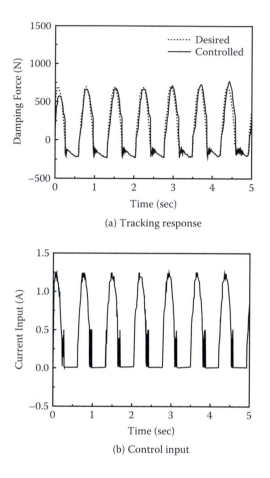

(a) Tracking response

(b) Control input

FIGURE 2.6
Damping force control response using a skyhook controller. (From Choi, S.B. et al., *Mechatronics*, 8, 2, 1998. With permission.)

input response that the required control input determined from the skyhook controller is applied during the rebound motion, while no magnetic field is applied during the jounce motion.

2.3 PID Control

An attractive controller to achieve desired damping force using MR damper system is a proportional-integral-derivative (PID) controller. As is well known, the PID controller is easy to implement in practice, and is very effective with robustness to system uncertainties. The control action of each P,

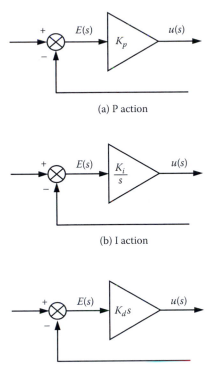

FIGURE 2.7
Control action of P, I, and D components.

I, and D is shown in Figure 2.7 [6]. From the block diagram, the input is expressed by

$$u(s) = k_p E(s), \text{ for } P \text{ action}$$

$$u(s) = \frac{k_i}{s} E(s), \text{ for } I \text{ action} \qquad (2.6)$$

$$u(s) = k_d s E(s), \text{ for } D \text{ action}$$

In the above, s is the Laplace variable. k_p, k_i, and k_d are control gains for P, I, and D components, respectively. $E(s)$ is the feedback error signal between desired value and actual output value. Consequently, the form of PID controller is given by

$$u(s) = k_p E(s) + \frac{k_i}{s} E(s) + k_d s E(s) \qquad (2.7)$$

The P controller is essentially an amplifier with an adjustable gain of k_p. If the k_p is increased, the response time of the control system becomes faster. But instability of the control system may occur using very high feedback gains of k_p. The value of the control $u(t)$ is changed at a rate proportional to the actuating error signal $e(t)$ by employing the I controller. For zero actuating error, the value of $u(t)$ remains to be stationary. By employing the I controller action, the steady state error of control system can be effectively alleviated or eliminated. This is a very significant factor to be considered in the tracking control problem. In general, we can increase system stability by employing the D controller. However, the D control action may amplify noise signals and cause a saturation effect in the MR actuator. It is also noted that the D control action can never be implemented alone because the control action is effective only during transient periods [6]. An appropriate determination of control gains k_p, k_i, and k_d to achieve superior control performance can be realized by several methods—Ziegler-Nichols, adaptive, and optimal.

In order to demonstrate the effectiveness of the PID controller, the MR damper for passenger vehicles is adopted and its damping force is controlled [7]. Figure 2.8 presents a block diagram to achieve the desired damping force of an MR damper using the PID controller. The desired damping force, F_d, to be tracked is stored in the microprocessor and compared with the actual damping force, F_a, generated from the MR damper. The actuating error signal is fed back to the PID controller, and hence control input is written by

$$u(t) = k_p e(t) + k_i \int_0^t e(t) + k_d \frac{de(t)}{dt}, \; e(t) = F_d - F_a \qquad (2.8)$$

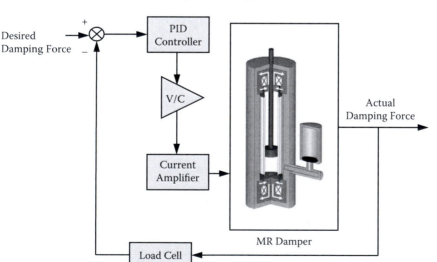

FIGURE 2.8
PID control scheme for damping force control of an MR damper.

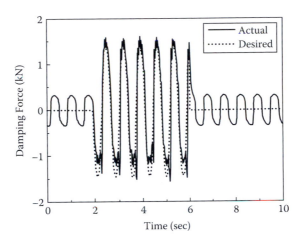

FIGURE 2.9

Damping force control response using a PID controller. (From Lee, H.S. et al., *Journal of Intelligent Material Systems and Structures*, 11, 80, 2000. With permission.)

The control input (current or magnetic field in this case) is then amplified via the voltage/current (v/c) converter. Figure 2.9 presents one of the control results obtained by implementing the PID controller [7]. It is clearly observed that the actual damping force from the MR damper tracks well to the desired one by applying control input current, which is determined by the PID controller. This result directly indicates that the damping force of the MR damper can be continuously controlled by the input current associated with the PID controller.

2.4 LQ Control

The linear quadratic (LQ) control is one of most popular control techniques that can be applicable to many control systems including MR actuator-based control systems. In this control method, the plant is assumed to be a linear system in the state space form and the performance index is a quadratic function of the plant states and control inputs. One of salient advantages of the LQ control method is that it leads to linear control laws that are easy to implement and analyze. For the linear quadratic regulator (LQR) type optimal control, the following state equation is considered [8].

$$\dot{x} = Ax + Bu \qquad (2.9)$$

where x is the state vector, u is the input vector, A is the system matrix, and B is the input matrix. The impending problem is to determine the optimal control vector

$$u(t) = -Kx(t) \tag{2.10}$$

so as to minimize the performance index

$$J = \int_0^\infty (x^T Q x + u^T R u) dt \tag{2.11}$$

where Q is the state weighting matrix (positive-semidefinite) and R is the input weighting matrix (positive-definite). The matrices Q and R determine the relative importance of the error and the expenditure of the control energy. If (A,B) is controllable, the feedback control gain is obtained by

$$K = R^{-1} B^T P \tag{2.12}$$

where P is the solution of the following algebraic Riccati equation:

$$A^T P + PA - PBR^{-1} B^T P + Q = 0 \tag{2.13}$$

If the performance index is given in terms of the output vector rather than the state vector, that is,

$$J = \int_0^\infty (y^T Q y + u^T R u) dt \tag{2.14}$$

then the index can be modified by using the output equation

$$y = Cx \tag{2.15}$$

to

$$J = \int_0^\infty (x^T C^T Q C x + u^T R u) dt \tag{2.16}$$

The design step to obtain the feedback gain K, which minimizes the index in Equation (2.16), is the same as the step for the feedback gain K, which minimizes the index in Equation (2.11). The LQ optimal control can be easily extended to the linear quadratic Gaussian (LQG) problem if the control system and the performance index are associated with white Gaussian noise as [8]:

$$\dot{x} = Ax + Bu + \Gamma \omega$$
$$y = Cx + v \tag{2.17}$$

$$\lim_{T \to \infty} \frac{1}{2T} E \left\{ \int_{-T}^{T} (x^T Q x + u^T R u) dt \right\} \tag{2.18}$$

where ω stands for random noise disturbance and v represents random measurement (sensor) noise. Both ω and v are white Gaussian zero-mean stationary processes. It is noted that because the states and control are both random, the performance index will be random. Thus, the problem is to find the optimal control that will minimize the average cost. Using the same procedure as for the LQR problem, the solution is achieved as:

  i. Controller:

$$u = -K\tilde{x}$$

$$K = R^{-1}B^T P \qquad (2.19)$$

$$A^T P + PA - PBR^{-1}B^T P + Q = 0$$

 ii. Estimator:

$$\dot{\tilde{x}} = A\tilde{x} + Bu + K_e(y - C\tilde{x})$$

$$K_e = P_1 C^T R_1^{-1} \qquad (2.20)$$

$$AP_1 + P_1 A^T - P_1 C^T R_1^{-1} C P_1 + \Gamma Q_1 \Gamma^T = 0$$

In the previous equations, Q_1 is the positive semi-definite matrix and R_1 is the positive definite matrix. It is noted that the problem can be solved in two separate stages—controller gain K and estimator gain K_e.

2.5 Sliding Mode Control

Despite many advantages of the feedback control systems, there exist some system perturbations (uncertainties) associated with MR devices. For instance, the field-dependent yield shear stress of MR fluid is subjected to change according to temperature variation. Moreover, the dynamic behavior of an MR device is a function of the magnetic field. There may exist non-linear hysteresis of the damping force in an MR damper system. Therefore, in order to guarantee control robustness of the control system featuring an MR device, a robust controller needs to be implemented to take account for system uncertainties.

A sliding mode controller (SMC), also called a variable structure controller, is well known as one of the most attractive candidates to assume control robustness against system uncertainties and external disturbances. The SMC has its roots in the literature of the former Soviet Union [9]. Today, throughout the world, the research and development on the SMC continue to apply it to a wide variety of engineering systems [10, 11]. Sliding modes that can be obtained by appropriate discontinuous control laws are the principal operation modes in the variable structure systems. The systems have invariance properties to the parameter variations and external disturbances under the sliding mode motion. In order to demonstrate the invariance property under the sliding mode motion, consider the following second-order system.

$$\dot{x}_1 = x_2$$
$$\dot{x}_2 = cx_2 - kx_1, \qquad c > 0, \text{ or } k < 0 \tag{2.21}$$

when $k < 0$, the eigenvalues of the system become $\lambda_{1,2} = \dfrac{c}{2} \pm \sqrt{\dfrac{c^2}{4} - k}$. Therefore,the phase portrait of the system is a saddle showing unstable motion except stable eigenvalue linear (refer to Figure 2.10a). When $k > 0$, the eigenvalues become $\lambda_{1,2} = \dfrac{c}{2} \pm \sqrt{\dfrac{c^2}{4} - k}$. Thus, the phase portrait of the system is a spiral source showing unstable motion (refer to Figure 2.10b). If the switching occurs on the line $s_g = cx_1 + x_2 = 0$ and on $x = 0$ with the following switching logic,

$$k = \begin{cases} negative, & x_1 s_g < 0; \quad (I) \\ positive, & x_1 s_g > 0; \quad (II) \end{cases} \tag{2.22}$$

then the system becomes asymptotically stable for any arbitrary initial conditions as shown in Figure 2.10c. Two subsystems converge to a line s_g (called switching line [surface] or sliding line [surface]). Once hitting the sliding line, the system can be described by

$$s_g = cx_1 + x_2 = 0: \quad \text{sliding mode equation} \tag{2.23}$$

This implies that the original system response is independent of system parameters on the sliding line (sliding mode motion). This guarantees the robustness of the system to system uncertainties and external disturbances. In general, the sliding mode motion can be achieved by satisfying the following so-called sliding mode condition [9]:

$$s_g \dot{s}_g < 0 \tag{2.24}$$

The above condition can be interpreted as the condition for Lyapunov stability [10]. In order to provide design steps for the SMC, consider the following control system subjected to the external disturbance:

$$\dot{x}_1 = x_2$$
$$\dot{x}_2 = ax_1 + x_2 + u + d \tag{2.25}$$

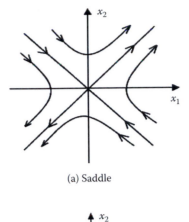

(a) Saddle

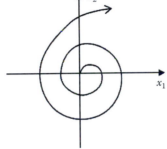

(b) Spiral source

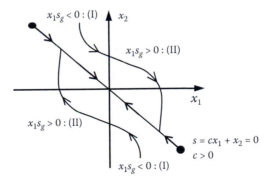

(c) With switching logic

FIGURE 2.10
Invariance property of the SMC.

where d is external disturbance and a is parameter variation. These are bounded by

$$|d| \leq \varepsilon, \quad a_1 \leq a \leq a_2 \tag{2.26}$$

As a first step, we choose a stable sliding line as

$$s_g = cx_1 + x_2 = 0, \quad c > 0 \tag{2.27}$$

Then, the sliding mode dynamics becomes

$$\dot{s}_g = cx_2 + ax_1 + x_2 + u \tag{2.28}$$

Thus, if we design the SMC, u, by

$$u = -cx_2 - x_2 - a_0 x_1 - (k + |a_m||x_1|) \operatorname{sgn}(s_g)$$

$$k > \varepsilon, \quad a_0 = (a_1 + a_2)/2, \quad a_m = a_2 - a_0 \tag{2.29}$$

The sliding mode condition in Equation (2.33) can be satisfied as:

$$s_g \dot{s}_g = (a - a_0) x_1 s_g - (k + |a_m||x_1|)|s_g| < 0 \tag{2.30}$$

In the controller given by Equation (2.38), k is the discontinuous control gain and sgn($\cdot$) is the signum function. This design step can be easily extended to higher order control systems [11].

References

[1] Leitmann, G. 1994. Semiactive control for vibration attenuation. *Journal of Intelligent Material Systems and Structures* 5: 841–846.

[2] Karnopp, D. 1974. Vibration control using semi-active force generators. *ASME Journal of Engineering for Industry* 96: 619–626.

[3] Guclu, R. 2005. Fuzzy logic control of seat vibrations of a non-linear full vehicle model. *Nonlinear Dynamics* 40: 21–34.

[4] Terano, T., Asai, K., and Sugeno, M. 1987. *Fuzzy System Theory and Its Applications*, Boston: Harcourt-Brace.

[5] Ogata, K. 1990. *Modern Control Engineering*, Englewood Cliffs, NJ: Prentice-Hall.

[6] Han, Y. M., Nam, M. H., Han, S. S., Lee, H. G., and Choi, S. B. 2002. Vibration control evaluation of a commercial vehicle featuring MR seat damper. *Journal of Intelligent Material Systems and Structures* 13: 575–579.

[7] Lee, H. S. and Choi, S. B. 2000. Control and response characteristics of a magnetorheological fluid damper for passenger vehicles. *Journal of Intelligent Material Systems and Structures* 11: 80–87.

[8] Chen, C. T. 1999. *Linear System Theory and Design*, New York: Oxford University Press.

[9] Zinober, A. S. I. 1990. Deterministic Control of Uncertain Systems, London: Peter Peregrinus Press.

[10] Slotine, J. J. E. and Sastry, S. S. 1983. Tracking control of nonlinear systems using sliding surfaces with application to robot manipulators, *International Journal of Control*, 38: 465–492.

[11] Park, D. W. and Choi, S. B. 1999. Moving sliding surfaces for high-order variable structure systems. *International Journal of Control* 72: 960–970.

3

Hysteretic Behaviors of Magnetorheological (MR) Fluid

3.1 Introduction

Hysteretic behavior is inherently encountered in many application devices that usually involve magnetic, ferroelectric, and mechanical systems. Smart materials such as magnetorheological (MR) fluids, electrorheological (ER) fluids, piezoceramics (PZTs), and shape memory alloys (SMAs) have recently been adopted as actuators for various active and semi-active control systems. These smart materials also exhibit nonlinear hysteretic responses, which have an adverse effect on actuator performance within control systems. Due to the increasing demand for superior control performance, nonlinear hysteresis models [1, 2] and robust feedback control schemes [3] have been intensively investigated by many researchers to improve or compensate for the hysteretic actuator behavior of smart materials.

Among the smart materials, MR fluid is actively being researched as an actuating fluid for valve systems, shock absorbers, engine mounts, and other control systems. As previously mentioned, MR fluid, which is reversibly and instantaneously changed by the application of a magnetic field to a fluid domain, is generally modeled as Bingham fluid [4, 5], which shows perfect Newtonian behavior after a yield point. However, MR fluid exhibits hysteretic behavior caused by changes in a dynamic condition (shear rate) and control input (magnetic field). Shear rate hysteresis has been widely studied because it can be easily considered with application device hysteresis, leading to many investigations on the hysteretic behavior of MR devices. Stanway et al. proposed an idealized mechanical model for ER damper behavior [6]. The mechanical model consisted of a Coulomb friction element placed in parallel with a viscous damper. Spencer et al. proposed a Bouc-Wen model [7] to describe MR damper behavior. The Bouc-Wen model parameters were estimated by fitting the damper's force-displacement and force-velocity behavior. Kamath and Wereley presented a nonlinear viscoelastic-plastic model [8] that accounts for the behavior of ER fluids in both the pre-yield and post-yield regimes for the applied shear. Wereley et al. also proposed a nonlinear biviscous model [9] for MR/ER dampers.

This chapter discusses new hysteresis modeling techniques for MR fluid, which can be easily integrated within a control system. Section 3.2 presents a key issue in MR device control by identifying a general hysteresis model of the MR fluid itself with respect to an applied magnetic field [10]. The conventional hysteresis models show a favorable approximation to describing the post-yield behavior of MR devices. However, we need more accurate models that can capture the complicated hysteretic behavior of MR fluids [7, 8, 11]. Furthermore, in the application devices, a magnetic field provides the control input to be applied to the MR fluid domain. Therefore, hysteretic behavior with respect to a control input can significantly affect application device control performance. In this section, the classical Preisach independent domain model, which is well known in the ferromagnetic field, is introduced to capture these characteristics of MR fluids. It has been adopted as an effective hysteresis model of PZT [12] and SMA [13]. In the test, the field-dependent shear stress of the commercial MR product (MRF-132LD; Lord Corporation, U.S.) is obtained using a rheometer with the MR cell. Two significant properties, the minor loop property and the wiping-out property [14, 15], are examined to show that the Preisach model is well matched to the physical hysteresis phenomenon of the MR fluid itself. A Preisach model for the MR fluid is then presented and numerically identified. Several first order descending (FOD) curves for the adopted MR fluid are experimentally derived to determine the relationship between shear stress and applied magnetic field. The identified Preisach model is verified via comparison of predicted shear stress with measured shear stress. The predicted shear stress is also compared between the proposed Preisach model and the conventional Bingham model.

Section 3.3 presents a polynomial hysteresis model for MR devices in which the polynomial expression has a variable control input coefficient that is used to easily control implementation [16]. Many types of semi-active MR dampers have recently been proposed to attenuate the vibrations of various dynamic systems including vehicle suspensions. Experimental realization has demonstrated that the unwanted vibrations of application systems can be electively controlled by employing MR dampers associated with appropriate control strategies [17, 18]. The use of an accurate damping force model, which can capture the inherent hysteretic behavior of MR dampers, is a very important factor in the successful achievement of desirable control performance. In particular, a more accurate damper model is required for realization of open-loop control, which is easier to implement and more cost effective than closed-loop control.

To date, several damper models have been proposed to predict the field-dependent hysteretic behavior of MR dampers: the Bouc-Wen model [7], the non-linear hysteretic biviscous model [9], and a modified Bingham plastic model [19]. The validity of these models for predicting the hysteretic behavior has been proven favorably by comparison with experimental results.

However, using these models, it is very difficult to realize a control system (open-loop or closed-loop) to achieve desirable tracking control performance of the field-dependent damping force because the experimental parameters used in the models are varied with respect to applied field intensity.

In this section, a cylindrical type of MR damper, which can be applied to a mid-sized passenger vehicle, is adopted and its hysteretic behavior is experimentally evaluated in the damping force versus piston velocity domain. The measured hysteretic characteristics of the field-dependent damping forces are compared with those predicted by the models listed before. The damping force control accuracy using the proposed model is also experimentally demonstrated using the open-loop control scheme.

3.2 Preisach Hysteresis Model Identification

3.2.1 Hysteresis Phenomenon

As we saw in the previous chapter, the MR fluid is a suspension consisting of magnetizable particles in a low permeability base fluid. The instantaneous change of rheological properties is caused by a polarization induced in the suspended particles. This phenomenon is associated with a yield stress of the suspension, which is subject to a magnetic field. MR fluids, in general, have been considered as the Bingham fluid whose constitutive equation is given by [4, 5]

$$\tau = \eta\dot{\gamma} + \tau_y(H) \tag{3.1}$$

where τ is the shear stress, η is the dynamic viscosity, $\dot{\gamma}$ is the shear rate, and $\tau_y(H)$ is the dynamic yield stress of the MR fluid. Generally, it has been proved that τ_y is a function of the magnetic field H. The Bingham model is widely used to control MR devices by the advantage of a favorable approximation to describe the post-yield behavior of an MR device. It is still incapable of describing the hysteretic behavior of an MR fluid itself.

Figure 3.1 represents an experimental apparatus to achieve a shear stress produced by the MR fluid. The field-dependent shear stress was measured using the rheometer (MCR 300, Physica) with the MR cell (TEK 70MR, Physica). An air bearing is adopted to minimize hysteresis effect caused by mechanical friction, and the operating temperature is kept at 25°C. The diameter of a rotating disk is 20 mm. The measuring gap between the rotating and the stationary disks is 1 mm, which contains the MR fluid. The personal computer associated with A/D (analog/digital) and D/A (digital/analog) converters controls both the shear rate and the input field.

Figure 3.2 shows the measured hysteretic behaviors of the MR fluid (MRF-132LD, Lord Corporation) adopted in this test. Two different hysteretic

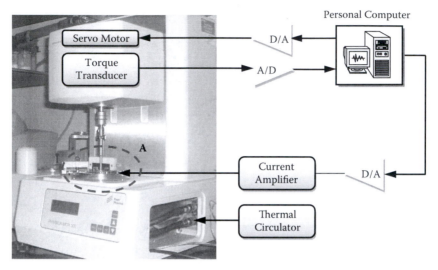

(a) Experimental apparatus

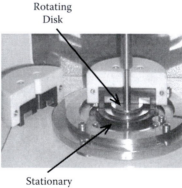

(b) Detail of A

FIGURE 3.1
Schematic configurations of experimental apparatus.

behaviors are measured because shear stress in Equation (3.1) is a function of both dynamic condition (shear rate) and control input (magnetic field). The shear stress shown in Figure 3.2(a) is measured under change of the shear rate while a constant magnetic field is applied. The shear stress shown in Figure 3.2(b) is measured under change of the magnetic field. The measured hysteresis loop is traced in the counterclockwise direction. It is clearly observed that there are hysteresis loops with respect to the shear rate and the intensity of the magnetic field. It is also shown that the effect of the magnetic field on the hysteretic behavior is much more significant

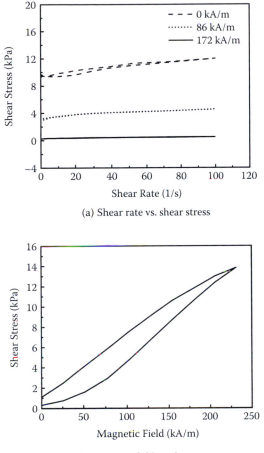

(a) Shear rate vs. shear stress

(b) Magnetic field vs. shear stress

FIGURE 3.2
Hysteretic behavior of the MR fluid. (From Han, Y.M. et al., *Journal of Intelligent Material Systems and Structures*, 19, 9, 2007. With permission.)

than the effect of the shear rate. In the application devices, the magnetic field acts as a control input. Therefore, the large hysteretic behavior with respect to the control input can significantly affect the control performance of application devices.

In this section, the measured shear stress at the shear rate of $1s^{-1}$ is considered an approximated yield stress in order to establish the hysteresis model between the yield stress (output) and magnetic field (input). By measuring the shear stress at the constant shear rate of $1s^{-1}$, it can be sufficiently assumed to be close to the dynamic yield stress as shown in Figure 3.3(a), while the field-dependent viscosity is kept below 50Pas as shown in Figure 3.3(b). It is remarked that the ordinates of figures titled by shear stress represent the dynamic yield stress.

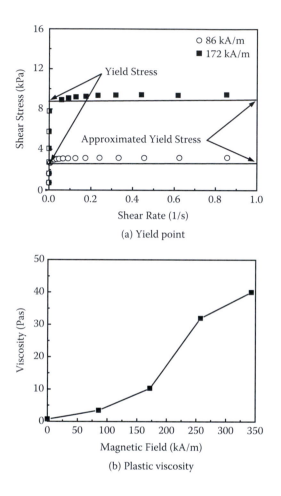

FIGURE 3.3

Approximated yield stress of the MR fluid. (From Han, Y.M. et al., *Journal of Intelligent Material Systems and Structures*, 19, 9, 2007. With permission.)

The Preisach model has been developed for modeling hysteresis in ferromagnetic materials. This model is a promising candidate to capture the hysteretic behavior of an MR fluid because the chain clusters of an MR fluid are formed by induced magnetic dipoles via electomagnetization. The Preisach model is phenomenological in nature, and it is characterized by two significant properties: the minor loop property and the wiping-out property [14, 15]. The minor loop property specifies that two comparable minor loops, which are generated by moving between two same pairs of input maximum and minimum, are to be congruent if one exactly overlaps the other after some shift in the output parameter. The wiping-out property specifies which values of the preceding input trajectory affect the current output, that is, which dominant maximum and minimum can wipe out the effects of preceding

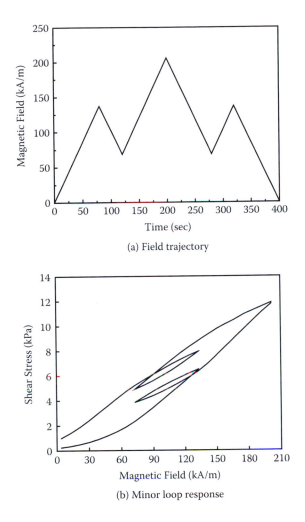

(a) Field trajectory

(b) Minor loop response

FIGURE 3.4
Minor loop property of the MR fluid. (From Han, Y.M. et al., *Journal of Intelligent Material Systems and Structures*, 19, 9, 2007. With permission.)

smaller dominant maxima and minima. In order to determine whether a physical hysteresis phenomenon is matched by the Preisach model, one can examine both the minor loop and the wiping-out properties.

Figure 3.4(a) shows the applied magnetic field trajectory for the minor loop experiment. The imposed trajectory is composed of triangular input signals. The achieved shear stress is shown in Figure 3.4(b). A favorable congruency is seen from the results. Figure 3.5(a) shows the applied magnetic field trajectory for the wiping-out experiment, which is composed of two sets of dominant maxima and minima. Figure 3.5(b) shows a good agreement between the first set of dominant maxima (a_1, a_2, a_3) and the second set ($\bar{a}_1, \bar{a}_2, \bar{a}_3$).

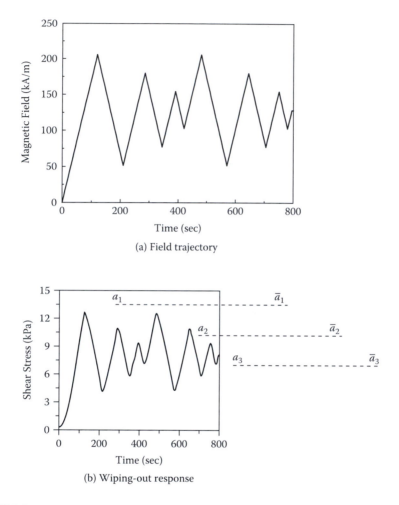

FIGURE 3.5
Wiping-out property of the MR fluid. (From Han, Y.M. et al., *Journal of Intelligent Material Systems and Structures*, 19, 9, 2007. With permission.)

This implies that the effect of the preceding former (a_1, a_2, a_3) has been wiped out by the larger maxima $(\bar{a}_1)$ of the latter $(\bar{a}_1, \bar{a}_2, \bar{a}_3)$, and thus the preceding input values (a_1, a_2, a_3) do not affect current outputs. Therefore, it is clearly seen that the wiping-out property is sufficiently satisfied for the MR fluid used in this test.

The results shown in Figure 3.4 and Figure 3.5 demonstrate a favorable congruency of the comparable minor loops and a satisfactory wiping-out property for the MR fluid. Thus, it is obvious that the Preisach model is applicable to describe the hysteresis behavior of the MR fluid. Based on these results, the hysteresis identification of the MR fluid was performed using the Preisach model.

3.2.2 Preisach Model

The Preisach model that we adopt to describe the hysteretic behavior of the MR fluid can be expressed as [14]:

$$\tau_y(t) = \iint_\Gamma \mu(\alpha,\beta)\gamma_{\alpha\beta}[H(t)]d\alpha\,d\beta \tag{3.2}$$

where Γ is the Preisach plane, $\gamma_{\alpha\beta}[\cdot]$ is the hysteresis relay, $H(t)$ is the magnetic field, $\tau_y(t)$ is the yield stress (or shear stress), and $\mu(\alpha,\beta)$ is the weighting function that describes the relative contribution of each relay to the overall hysteresis. Each relay is characterized by the pair of switching values (α,β) with $\alpha \geq \beta$. As the input varies with time, each individual relay adjusts its output according to the magnetic field input value, and the weighted sum of all relay outputs provides the overall system output as shown in Figure 3.6(a). The simplest possible hysteresis relay for the MR fluid is shown in Figure 3.6(b). It is a modification of a classical relay that has two states, –1 and 1, corresponding to the opposite polarization of a ferromagnetic material. The output of the adopted relay is either 0 or 1. This is suitable for modeling hysteresis of the MR fluid where the output (yield stress) varies between zero and a maximum value. In this case, a hysteresis loop is located in the first quadrant of the input-output (magnetic field-yield stress) plane. The Preisach plane can be geometrically interpreted as one-to-one mapping between relays and switching values of (α,β) as shown in Figure 3.6(c). The maximum yield stress restricts the plane within the triangle in Figure 3.6(c), where α_0 and β_0 represent the upper and lower limits of the magnetic field input. In this test, they are chosen by 0 kA/m and 257.25 kA/m, respectively. The Preisach plane provides the state of individual relay, and thus the plane is divided into two time-varying regions as:

$$\Gamma_- = \{(\alpha,\beta) \in \Gamma \mid \text{output of } \gamma_{\alpha\beta} \text{ is } 0\}$$
$$\Gamma_+ = \{(\alpha,\beta) \in \Gamma \mid \text{output of } \gamma_{\alpha\beta} \text{ is } 1\} \tag{3.3}$$

The two regions represent that relays are on 0 and 1 positions, respectively. Therefore, Equation (3.2) can be reduced to

$$\tau_y(t) = \iint_{\Gamma_+} \mu(\alpha,\beta)d\alpha d\beta \tag{3.4}$$

The use of a numerical technique for the Preisach model identification has been normally proved as an effective way for smart materials [1, 12]. Therefore, the Preisach model for the MR fluid is numerically implemented by employing a numerical function for each mesh value of (α,β) within the Preisach plane. The mesh values are identical with the data in the FOD curves,

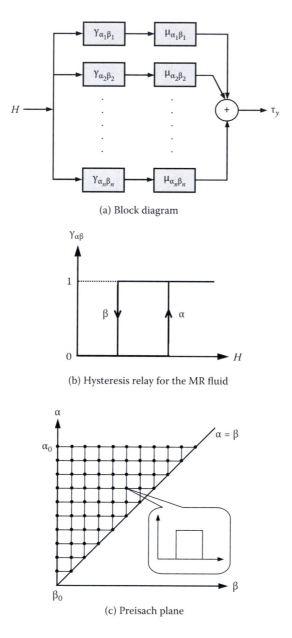

(a) Block diagram

(b) Hysteresis relay for the MR fluid

(c) Preisach plane

FIGURE 3.6
Configuration of the Preisach model.

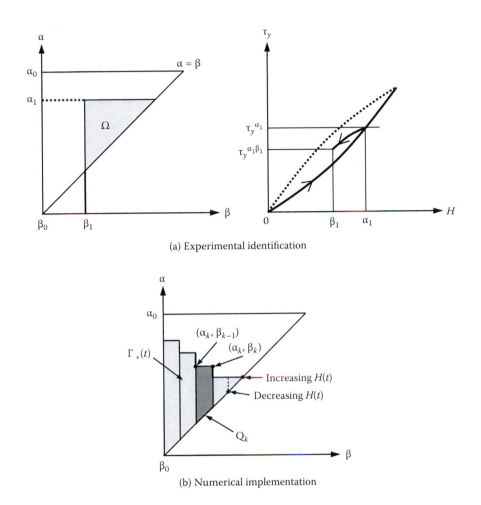

(a) Experimental identification

(b) Numerical implementation

FIGURE 3.7
Numerical identification and implementation of the Preisach model.

which are experimentally obtained. Figure 3.7(a) shows one of the mesh values of (α_1, β_1), and its corresponding FOD curve is a monotonic increase to a value α_1, then a monotonic decrease to β_1. After the input peaks at α_1, the decrease sweeps out area Ω, generating the descending branch inside the major loop. A numerical function $T(\alpha_1, \beta_1)$ is then defined as the output change along the descending branch as:

$$T(\alpha_1, \beta_1) = \iint_{\Omega} \mu(\alpha, \beta)\, d\alpha\, d\beta$$

$$= \tau_y^{\alpha_1} - \tau_y^{\alpha_1\beta_1}$$

(3.5)

From Equation (3.4) and Equation (3.5), we can determine an explicit formula for the hysteresis in terms of experimental data.

Figure 3.7(b) shows the increasing and decreasing series of an input magnetic field, where Γ_+ can be subdivided into n trapezoids Q_k. Geometrically, the area of each trapezoid Q_k can be expressed as a difference of two triangle areas concerned with $T(\alpha_k, \beta_{k-1})$ and $T(\alpha_k, \beta_k)$, respectively. Therefore, the output for each Q_k is derived by

$$\iint\limits_{Q_k} \mu(\alpha,\beta)\,d\alpha\,d\beta = T(\alpha_k,\beta_{k-1}) - T(\alpha_k,\beta_k) \tag{3.6}$$

By summing the area of trapezoids Q_k over the entire area Γ_+, the output is expressed by

$$\tau_y(t) = \sum_{k=1}^{n(t)} \iint\limits_{Q_k} \mu(\alpha,\beta)\,d\alpha\,d\beta$$

$$= \sum_{k=1}^{n(t)} \left[T(\alpha_k,\beta_{k-1}) - T(\alpha_k,\beta_k) \right] \tag{3.7}$$

Consequently, for the cases of the increasing and decreasing input, the output $\tau_y(t)$ of the Preisach model is expressed by the experimentally defined $T(\alpha_k, \beta_k)$ [1, 12]:

$$\tau_y(t) = \sum_{k=1}^{n(t)-1} \left[T(\alpha_k,\beta_{k-1}) - T(\alpha_k,\beta_k) \right] + T(H(t),\beta_{n(t)-1}), \quad \text{for increasing input}$$

$$\tau_y(t) = \sum_{k=1}^{n(t)-1} \left[T(\alpha_k,\beta_{k-1}) - T(\alpha_k,\beta_k) \right] + \left[T(\alpha_{n(t)},\beta_{n(t)-1}) - T(\alpha_{n(t)},H(t)) \right], \tag{3.8}$$

$$\text{for decreasing input}$$

Numerical implementation of the Preisach model requires experimental determination of $T(\alpha_k, \beta_k)$ at a finite number of grid points within the Preisach plane. Practically, the finite number of grid points and the corresponding measured output values cause a certain problem that some values of the magnetic field input do not lie at the grid point. Furthermore, it may not have the measured output values. In this work, the following interpolation functions are employed to determine the corresponding $T(\alpha_k, \beta_k)$ when the value of a magnetic field input does not lie at the grid point:

$$\tau_y^{\alpha_k \beta_k} = c_0 + c_1 \alpha_k + c_2 \beta_k + c_3 \alpha_k \beta_k, \quad \text{for square cells}$$

$$\tau_y^{\alpha_k \beta_k} = c_0 + c_1 \alpha_k + c_2 \beta_k, \quad \text{for triangular cells} \tag{3.9}$$

3.2.3 Hysteresis Identification and Compensation

As a first step, FOD data sets are constructed to calculate the numerical function of the Preisach model. Figure 3.8(a) shows the magnetic field input to collect FOD curves. The range of the magnetic field is from 0 kA/m to 257.25 kV/m, and it is partitioned into 10 sub-ranges. The magnetic field is applied in a stepwise manner, and each step of the magnetic field is maintained for 20 s to satisfy

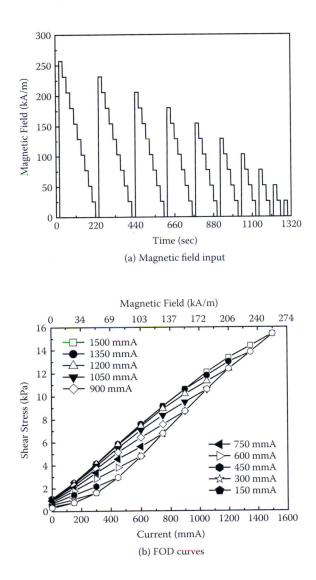

(a) Magnetic field input

(b) FOD curves

FIGURE 3.8
Measured FOD curves. (From Han, Y.M. et al., *Journal of Intelligent Material Systems and Structures*, 19, 9, 2007. With permission.)

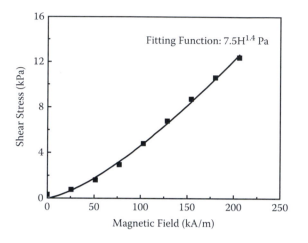

FIGURE 3.9

Bingham characteristics of the MR fluid. (From Han, Y.M. et al., *Journal of Intelligent Material Systems and Structures*, 19, 9, 2007. With permission.)

a steady state condition. From the measured shear stress using the rheometer, FOD curves were achieved as shown in Figure 3.8(b). Then, the Preisach model for the MR fluid is numerically identified by using the FOD data sets.

A hysteresis prediction using the Preisach model was tested under three different types of input trajectories of the magnetic field: step, triangular, and arbitrary signals. The predicted results were compared to the results predicted by the Bingham model in Equation (3.1). Figure 3.9 shows the Bingham characteristics of the MR fluid adopted in this test, from which the field-dependent yield stress is obtained by $\tau_y(H) = 7.5H^{1.4}$ Pa.

Figure 3.10(b) presents the actual and predicted hysteresis responses under a step input shown in Figure 3.10(a). In increasing step, the predicted responses by the Preisach model and the Bingham model show similar results, but the hysteresis effect shows distinctly in the next decreasing step. As clearly observed from the results, the Preisach model predicts the actual responses with great accuracy, and it shows much better accuracy than the Bingham model. It is verified from the prediction error shown in Figure 3.10(c). The maximum prediction errors are calculated by 0.72 kPa and 2.34 kPa in a steady state condition for the Preisach model and the Bingham model, respectively.

Figure 3.11 presents the hysteresis response under a triangular input trajectory. It is clearly seen that the hysteresis nonlinearity is well followed by the proposed Preisach model with a small prediction error as shown in Figure 3.11(c). The maximum prediction errors of Preisach and Bingham models are calculated by 0.43 kPa and 2.49 kPa, respectively. Figure 3.12 compares the field-dependent hysteresis loops between measurement (actual) and identification from the proposed Preisach model. It is clearly observed

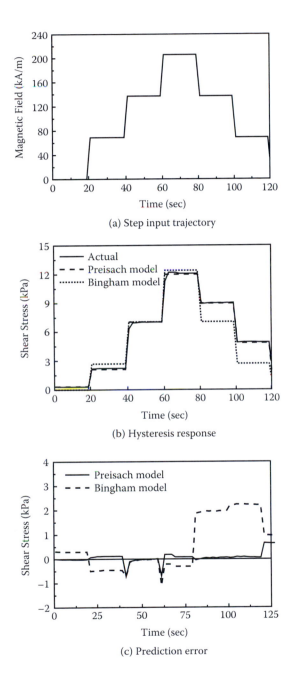

FIGURE 3.10
Actual and predicted hysteresis responses under a step input. (From Han, Y.M. et al., *Journal of Intelligent Material Systems and Structures*, 19, 9, 2007. With permission.)

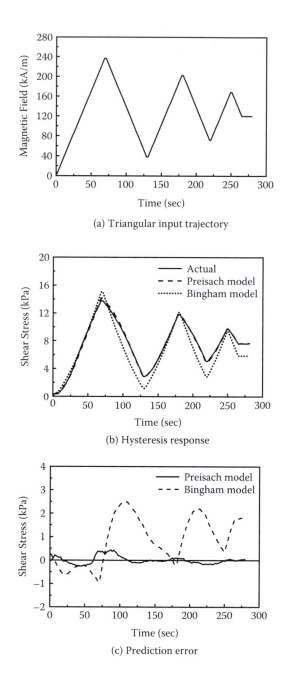

(a) Triangular input trajectory

(b) Hysteresis response

(c) Prediction error

FIGURE 3.11
Actual and predicted hysteresis responses under a triangular input. (From Han, Y.M. et al., *Journal of Intelligent Material Systems and Structures,* 19, 9, 2007. With permission.)

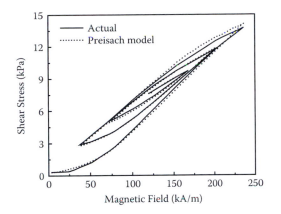

FIGURE 3.12
Hysteresis loops of the MR fluid. (From Han, Y.M. et al., *Journal of Intelligent Material Systems and Structures*, 19, 9, 2007. With permission.)

that the Preisach model well represents the field-dependent hysteretic behavior of the shear stress (yield stress) in the MR fluid.

Figure 3.13 presents the prediction responses under an arbitrary input trajectory, which is composed of different slopes of ramp and constant inputs. The Preisach model also predicts well the hysteresis output. From the results, it is obvious that the proposed Preisach model captures very well the hysteresis behavior of the MR fluid itself between output (yield stress) and input (magnetic field).

The control issue of MR devices is adjusting the magnetic field to achieve the desired shear (or yield) stress, which is concerned with an actuating force. One popular control strategy that is very effective for hysteresis nonlinearity is an inverse model control. Therefore, a simple open loop compensation strategy is achieved through Preisach model inversion, which is model-based compensation of the hysteresis nonlinearity. If the applied magnetic field is given for a known shear rate, a shear stress generated by the MR fluid can be predicted from the Preisach model. In contrast, the magnetic field for generating a desired shear stress can be calculated using the inverse model. In this simple open loop strategy, control performance is significantly affected by the accuracy of the formulated model. Figure 3.14 shows a flow chart of the proposed control algorithm. It is composed of predicting and linearizing the hysteresis nonlinearity in a discrete manner using the Preisach model. After specifying a set of desired shear stress trajectories, $\tau_d(k)$, a corresponding desired magnetic field, $H_d(k)$, is calculated using the nominal relationship between the magnetic field and shear stress. The nominal relationship can be obtained from the Bingham characteristic yield curve of the MR fluid. Then, the kth predicted shear stress is calculated from the kth desired magnetic field and the Preisach model. In order to linearize, the predicted shear stress, $\tau_d(k)$, is compared with the desired shear stress, $\tau_d(k)$, and then the algorithm

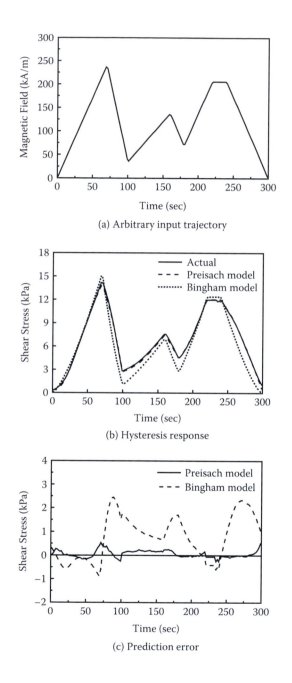

(a) Arbitrary input trajectory

(b) Hysteresis response

(c) Prediction error

FIGURE 3.13
Actual and predicted hysteresis responses under an arbitrary input. (From Han, Y.M. et al., *Journal of Intelligent Material Systems and Structures*, 19, 9, 2007. With permission.)

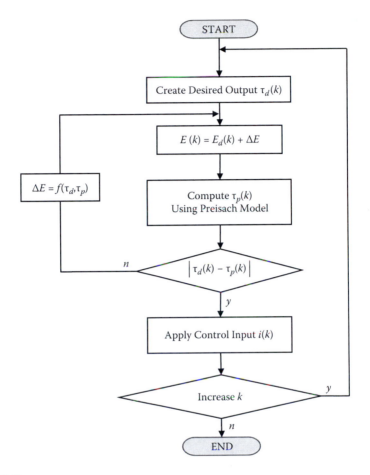

FIGURE 3.14
The proposed compensation strategy.

updates the real control input repeatedly until its error is sufficiently small. Therefore, the final *k*th control input *u(k)* can be written as:

$$u(k) = H_d(k) + \sum_{i=1}^{m} \left\{ \frac{\left[\tau_d(k) - \tau_p(k, K_c) \right]_i}{\alpha} \right\}^{\frac{1}{\beta}}, \quad \text{for } \left| \tau_d(k) - \tau_p(k, K_c) \right| < \varepsilon \quad (3.10)$$

where ε is the error bound and *m* is the number of updating times. The values of α and β are obtained from the Bingham characteristic yield curve of the MR fluid. After the *k*th control input is obtained, the next *k*+1th desired shear stress is introduced, and this process is repeated for the entire desired shear stress set. As shown here, the proposed hysteresis model of MR fluids is easily integrated into control systems.

3.3 Polynomial Hysteresis Model Identification

3.3.1 Hysteresis Phenomenon

The schematic configuration of an automotive MR damper adopted in the test is shown in Figure 3.15. The MR damper is divided into the upper and lower chambers by the piston, and it is fully filled with the MR fluid—Lord product MRF132-LD. The principal design parameters are chosen as follows—the outer radius of the inner cylinder: 30.1 mm; the length of the magnetic pole: 10 mm; the gap between the magnetic poles: 1.0 mm; the number of coil turns: 150; and the diameter of the copper coil: 0.8 mm. The gas chamber is fully charged by nitrogen and its initial pressure at the maximum extension (up motion of the piston) is set at 25 bar. Figure 3.16 presents the measured damping force versus piston velocity at various input currents (magnetic fields). The result is obtained by exciting the MR damper with the excitation frequency of 1 to 4 Hz and the exciting magnitude of ±20 mm. The details

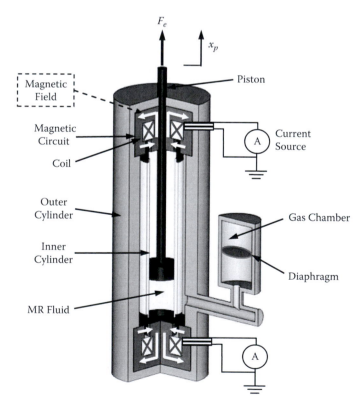

FIGURE 3.15
Configuration of the MR damper.

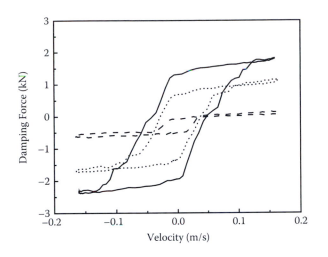

FIGURE 3.16
Measured damping force characteristics. (From Choi, S.B. et al., *Journal of Sound and Vibration*, 245, 2, 2001. With permission.)

for the measurement procedures are well described in Reference [20]. It is clearly observed from Figure 3.16 that the magnitude of the damping force at a certain piston velocity increases as the input current increases. Moreover, it is seen that the hysteresis loop is also increased with the increment of the input current.

3.3.2 Polynomial Model

Let us begin with conventional damper models to predict the damping force characteristics. Figure 3.16 shows the field-dependent damping force characteristics of the MR damper. The first model is a simple Bingham model and its basic mechanism is represented by Figure 3.17(a). In the Bingham model, the yield stress (τ_y) of the MR fluid is expressed by

$$\tau_y = \alpha H^\beta, \tag{3.11}$$

where H is the magnetic field, and α and β are intrinsic values of the employed MR fluid, which are to be experimentally identified. Thus, the damping force of the MR damper can be obtained [17] by

$$F = k_e x^p + c_e \upsilon + \alpha_1 \alpha H^\beta \text{sign}(\upsilon) \tag{3.12}$$

In Equation (3.12), x^p is the piston displacement, υ is the piston velocity, k_e is the stiffness constant due to the gas compliance, c_e is the damping constant due to the viscosity of the MR fluid, and α_1 is a geometrical constant.

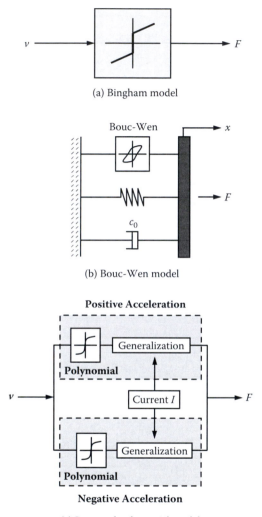

(a) Bingham model

(b) Bouc-Wen model

(c) Proposed polynomial model

FIGURE 3.17
Models for damping force prediction.

The third term is, of course, the controllable damping force by the input magnetic field (or current) of H. Figure 3.17(b) presents a basic mechanism of the Bouc-Wen model, which is frequently adopted in the analysis of non-linear hysteresis behavior. The damping force of the MR damper can be given [7] by

$$F = c_0 v + k_0(x - x_0) + \gamma z \qquad (3.13)$$

where x_0 is the initial displacement due to the gas, γ is the pressure drop due to the MR effect (yield stress), and z is obtained by

$$\dot{z} = -\varepsilon\,|v|z|z|^{n-1} - \delta v\,|z|^n + Av \tag{3.14}$$

In the above, ε, δ, and A are experimental parameters of the Bouc-Wen model that affect the hysteresis behavior in the pre-yield region. It is noted that experimental parameters of ε, δ, and A are varied with respect to the intensity of the input field. Therefore, it is very difficult to realize an open-loop control system to obtain a desirable damping force.

Figure 3.17(c) shows a schematic configuration of the third model focused on in this section. We can divide the hysteresis loop in Figure 3.16 into two regions: positive acceleration (lower loop) and negative acceleration (upper loop). Then, the lower loop or the upper loop can be fitted by the polynomial with the power of piston velocity. Therefore, the damping force of the MR damper can be expressed by

$$F = \sum_{i=0}^{n} a_i v^i, \quad n = 6 \tag{3.15}$$

where a_i is the experimental coefficient to be determined from the curve fitting. It is noted that the order of the polynomial can be chosen based on trial and error. It turned out that the polynomials up to the fifth order could not capture the measured hysteresis behavior. In addition, it has been observed that sixth and higher order polynomials favorably capture the hysteresis behavior without much difference. Therefore, a sixth order polynomial was chosen by considering computational time, which is also a very important factor in the real-time control of the damping force. The coefficient a_i in Equation (3.15) can be represented with respect to the intensity of the input current as shown in Figure 3.18. In the plots, the dark square indicates the measured value, and the solid curve is the linear fit of the coefficient a_i. The plots for a_1, a_3, a_4, a_5, and a_6 are referred to in Reference [21]. It is clearly observed that the coefficient a_i can be linearized with respect to the input current as:

$$a_i = b_i + c_i I, \quad i = 0, 1, \ldots, 6 \tag{3.16}$$

As a result, the damping force can be expressed by

$$F = \sum_{i=0}^{n} (b_i + c_i I) v^i \tag{3.17}$$

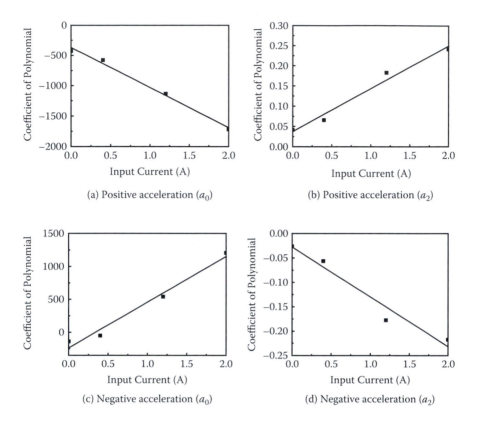

FIGURE 3.18
The relationship between a_0 and current. (From Choi, S.B. et al., *Journal of Sound and Vibration*, 245, 2, 2001. With permission.)

The coefficients b_i and c_i are obtained from the intercept and the slope of the plots shown in Figure 3.18. The specific values of b_i and c_i used in this test are listed in Table 3.1. It is noted that the coefficients a_i, b_i, and c_i are not sensitive to the magnitude of the input current [21].

Thus, we can easily realize an open-loop control system to achieve a desirable damping force. This is presented in section 3.3.3.

3.3.3 Hysteresis Identification and Compensation

The measured damping force is compared with the predicted damping forces obtained from the Bingham model, the Bouc-Wen model, and the proposed polynomial model as shown in Figure 3.19. The excitation frequency and magnitude are chosen as 1.4 Hz and ±20 mm, respectively. The input current applied to the MR damper is set as 1.2 A. It is clearly observed that the Bingham model cannot capture the non-linear hysteresis behavior, although it fairly predicts only the magnitude of the damping force at a certain piston

TABLE 3.1

Coefficients b_i and c_i of the Polynomial Model

Positive Acceleration			
Parameter	Value	Parameter	Value
b_0	−371.8	c_0	−659.4
b_1	6.205	c_1	8.955
b_2	0.03728	c_2	0.1062
b_3	−3.487e−4	c_3	−1.584e−4
b_4	−2.767e−6	c_4	−5.908e−6
b_5	6.924e−9	c_5	1.137e−9
b_6	5.604e−11	c_6	1.087e−10
Negative Acceleration			
Parameter	Value	Parameter	Value
b_0	−235.8	c_0	693.7
b_1	5.391	c_1	7.034
b_2	−0.02774	c_2	−0.1020
b_3	−3.788e−4	c_3	6.729e−5
b_4	2.449e−4	c_4	4.967e−6
b_5	8.804e−9	c_5	−4.924e−9
b_6	−5.374e−11	c_6	−8.196e−11

velocity. On the other hand, the measured hysteresis behavior is well predicted by the Bouc-Wen model or the polynomial model. In order to demonstrate the general effectiveness of the proposed model, we change the excitation conditions and the input current. The comparative results between the measurement and the simulation under various operating conditions are shown in Figure 3.20. We clearly see that the proposed model predicts fairly well the hysteresis behavior under various conditions without modifying the experimental coefficients of a_i, b_i, and c_i as mentioned earlier. An accuracy of damping force control of the MR damper depends upon the damper model. To demonstrate this, an open-loop control system to achieve a desirable damping force is established as shown in Figure 3.21. Once the desirable damping force is set in the microprocessor, the control input current to achieve the desirable damping force is determined from the damper model and applied to the MR damper. For the proposed damper model, the control input is determined from Equation (3.17) and is given by

$$I = \frac{F_d - \sum_{i=0}^{n} b_i v^i}{\sum_{i=0}^{n} c_i v^i} \tag{3.18}$$

where F_d is the desirable damping force to be tracked. The desirable damping force is normally set by $F_d = c_{sky} v$. The coefficient c_{sky} is control gain, and it is

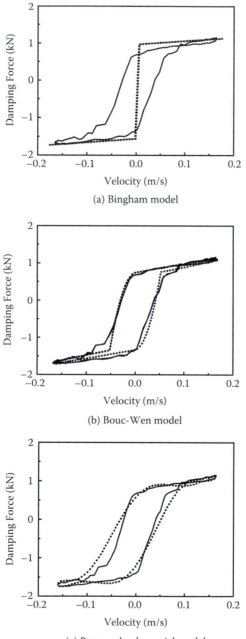

(a) Bingham model

(b) Bouc-Wen model

(c) Proposed polynomial model

FIGURE 3.19
Comparison of damping forces between the measurement and the prediction. (From Choi, S.B. et al., *Journal of Sound and Vibration*, 245, 2, 2001. With permission.)

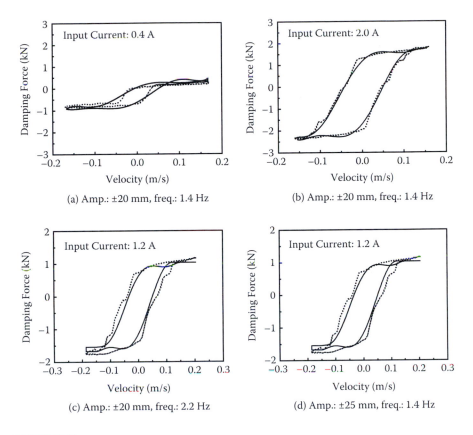

FIGURE 3.20
Damping force characteristics at various operating conditions. (From Choi, S.B. et al., *Journal of Sound and Vibration*, 245, 2, 2001. With permission.)

chosen as 13,000 in this test. Figure 3.22 presents the damping force control-lability realized from the open-loop control system. It is clearly seen that the control accuracy of the proposed model is much better than the Bingham model. In the Bingham model, it is observed that the tracking accuracy at the peak is not so good. This is because the Bingham model cannot capture the

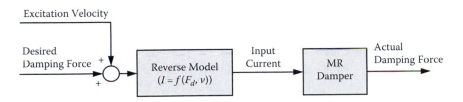

FIGURE 3.21
Block diagram for damping force control.

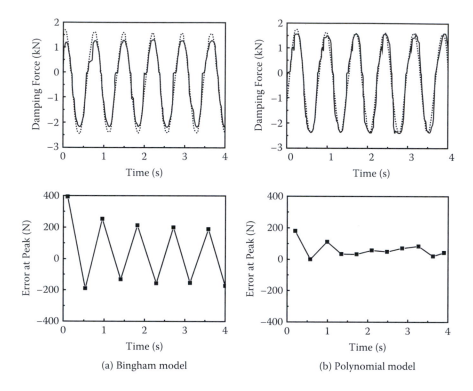

(a) Bingham model   (b) Polynomial model

FIGURE 3.22

Tracking control responses of the damping force. (From Choi, S.B. et al., *Journal of Sound and Vibration*, 245, 2, 2001. With permission.)

damping force behavior at zero and near-zero piston velocity as shown in Figure 3.19(a). On the contrary, the polynomial model tracks well the desired one in the whole range of the piston velocity.

3.4 Some Final Thoughts

In this chapter, two hysteresis modeling techniques are discussed as potential candidates, which can be easily integrated into a control system adopting MR fluids.

The first section identified a general hysteresis model of the MR fluid itself with respect to an applied magnetic field, which acts as a control input of MR devices. The field-dependent hysteretic behavior has been identified using the Preisach model. The applicability of the Preisach model to the MR fluid was demonstrated through the minor loop test and wiping-out test. The hysteresis model was then established using the first-order descending curves and numerical method. Its effectiveness was verified by comparing the pre-dicted field-dependent shear stress with the measured one. It has been dem-onstrated that the Preisach model can substantially reduce the prediction error under various input trajectories in comparison with the conventional Bingham model.

The second section discussed a hysteresis model using polynomial expres-sion for the field-dependent damping force of the MR damper. Its hysteretic damping force is compared with the predicted ones from the Bingham model and the Bouc-Wen model. It has been demonstrated that the polynomial model predicts fairly well the non-linear hysteresis behavior of the MR damper. In addition, the superior control accuracy of the proposed model to the Bingham model was verified by realizing the open-loop control system to track a desirable damping force.

It is finally remarked that the introduced hysteresis compensation meth-ods can be easily integrated into the control systems utilizing MR application devices such as electronic control suspension adopting MR shock absorbers. The hysteretic characteristics of the MR fluid are observed as a function of both dynamic condition (shear rate) and control input (magnetic field). In order to enhance control accuracy of MR applications, this complex and large hysteretic behavior should be considered in controller synthesis. The hyster-esis model of the MR fluid therefore becomes more significant in practical situations requiring easy implementation.

References

[1] Mittal, S. and Menq, C. H. 2000. Hysteresis compensation in electromag-netic actuators through Preisach model inversion. *IEEE/ASME Transactions on Mechatronics* 5: 394–409.

[2] Song, D. and Li, J. C. 1999. Modeling of piezoactuator's nonlinear and frequency dependent dynamics. *Mechatronics* 9: 391–410.

[3] Choi, S. B. and Lee, C. H. 1997. Force tracking control of a flexible gripper driven by a piezoceramic actuator. *ASME Journal of Dynamic Systems, Measurement and Control* 119: 439–446.

[4] Shames, I. H. and Cozzarelli, F. A. 1992. *Elastic and Inelastic Stress Analysis,* Upper Saddle River, NJ: Prentice Hall.

[5] Choi, S. B., Nam, M. H., and Lee, B. K. 2000. Vibration control of a MR seat damper for commercial vehicles. *Journal of Intelligent Material Systems and Structures* 11: 936–944.

[6] Stanway, R., Sproston, J. L., and Stevens, N. G. 1987. Non-linear modeling of an electro-rheological vibration damper. *Journal of Electrostatics* 20: 167–184.

[7] Spencer Jr., B. F., Dyke, S. J., Sain, M. K., and Carlson, J. D. 1997. Phenomenological model for a magnetorheological damper. *Journal of Engineering Mechanics, American Society of Civil Engineers* 230: 3–11.

[8] Kamath, G. M. and Wereley, N. M. 1997. Nonlinear viscoelastic-plastic mechanisms-based model of an electrorheological damper. *Journal of Guidance, Control and Dynamics* 20: 1125–1132.

[9] Wereley, N. M., Pang, L., and Kamath, G. M. 1998. Idealized hysteresis modeling of electrorheological and magnetorheological dampers, *Journal of Intelligent Material Systems and Structures* 9: 642–649.

[10] Han, Y. M., Choi, S. B., and Wereley, N. M. 2007. Hysteretic behavior of magnetorheological fluid and identification using Preisach model. *Journal of Intelligent Material Systems and Structures* 18: 973–981.

[11] Han, Y. M., Lim, S. C., Lee, H. G., Choi, S. B., and Choi, H. J. 2003. Hysteresis identification of polymethylaniline-based ER fluid using Preisach model. *Materials & Design* 24: 53–61.

[12] Ge, P. and Jouaneh, M. 1997. Generalized Preisach model for hysteresis nonlinearity of piezoceramic actuators, *Precision Engineering* 20: 99–111.

[13] Gorbet, R. B., Wang, D. W. L., and Morris, K. A., 1998. Preisach model identification of a two-wire SMA actuator. *Proceedings of the IEEE International Conference on Robotics & Automation*, Leuven, Belgium, pp. 2161–2167.

[14] Mayergoyz, I. D. 1991. *Mathematical Models of Hysteresis*. New York: Springer-Verlag.

[15] Hughes, D. and Wen, J. T. 1997. Preisach modeling of piezoceramic and shape memory alloy hysteresis. *Smart Materials and Structures* 6: 287–300.

[16] Choi, S. B. and Lee, S. K. 2001. A hysteresis model for the field-dependent damping force of a magnetorheological damper. *Journal of Sound and Vibration* 245: 375–383.

[17] Choi, S. B., Choi, Y. T., and Park, D. W. 2000. A sliding mode control of a full-car ER suspension via hardware-in the-loop-simulation. *Journal of Dynamic Systems and Measurement and Control* 122: 114–121.

[18] Carlson, J. D. and Sproston J. L. 2000. Controllable fluid in 2000-status of ER and MR fluid technology. *Seventh International Conference on New Actuators*, Bremen, Germany, pp. 126–130.

[19] Sims, N. D., Peel, D. J., Stanway, R., Johnson, A. R., and Bullough, W. A. 2000. The electrorheological long-stroke damper: a new modeling technique with experimental validation. *Journal of Sound and Vibration* 229: 207–227.

[20] Choi, S. B., Choi, Y. T., Chang, E. G., Han, S. J., and Kim, C. S. 1998. Control characteristics of a continuously variable ER damper. *Mechatronics* 8: 143–161.

[21] Lee, S. K. 2000. Hysteresis model of damping forces of MR damper for a passenger car, M.S. Thesis, Inha University, Korea.

4

Magnetorheological (MR) Suspension System for Passenger Vehicles

4.1 Introduction

Vehicle vibration from various road conditions requires attenuation. Successful suppression of vibration improves the lifespan of vehicle components, ride comfort, and steering stability. This is normally accomplished by a vehicle's suspension system, which generally consists of springs and shock absorbers (also called dampers). Various suspension methods have been developed and applied to control vehicle vibration. Vehicle suspension systems are classified into three types—passive, active, and semi-active—based on the amount of external power required by the system [1].

A passive suspension system featuring conventional dampers provides design simplicity and cost effectiveness, but performance limitation is inevitable due to uncontrollable damping forces. Thus, the passive system cannot provide superior vibration isolation under various road conditions. An active suspension system compensates for this limitation. With an additional active force and control algorithm introduced as a part of a suspension unit, the system is controlled to be more responsive to disturbances. An active suspension system generally provides high control performance in a wide frequency range, but it requires large power sources, complicated components such as sensors and servo valves, and a sophisticated control algorithm. A semi-active (also known as adaptive-passive) suspension configuration addresses these drawbacks by effectively integrating a control scheme with tunable devices. Thereby, active force generators are replaced with modulated variable compartments such as variable rate dampers. The semi-active suspension can offer desirable performance without requiring large power sources or expensive hardware.

Various semi-active shock absorbers featuring MR fluid have recently been proposed and successfully applied in vehicle suspension systems. Carlson et al. [1] proposed a commercially available MR damper that is applicable to on- and off-highway vehicle suspension systems. They experimentally demonstrated that sufficient levels of damping force and superior control capability of the damping force could be achieved by application of a control magnetic field. Spencer Jr. et al. [2] proposed a dynamic model for predicting the damping force of an MR damper and compared the measured damping forces with those predicted in the time domain. Kamath et al. [3] proposed a dynamic, semi-active MR lag mode damper model and verified its validity by comparing the predicted and measured damping forces. Yu et al. [4] evaluated the effective performance of the MR suspension system using a road test. Guo and Hu [5] proposed a nonlinear stiffness MR damper model that they verified using a simulation and an experiment.

Du, Sze, and Lam [6] proposed an H-infinity control algorithm for a vehicle MR damper and verified its effectiveness using a simulation. Shen, Golnaraghi, and Heppler [7] proposed a load-leveling suspension with an MR damper. Pranoto and Nagaya [8] proposed a 2DOF-type rotary MR damper and verified its efficiencies. Ok et al. [9] proposed cable-stayed bridges using MR dampers and verified their effectiveness using a semi-active fuzzy control algorithm. Choi et al. [10] manufactured an MR damper for a passenger vehicle and presented its control characteristics of the damping force. Based on this work, they extended their research and evaluated the control performance of the proposed MR damper using a hardware-in-the-loop simulation (HILS) [11].

This chapter discusses a semi-active suspension system featuring MR shock absorbers, whose subjects are categorized by optimal design, damping force control, and vibration control. Section 4.1 first details the optimal design of a semi-active shock absorber featuring MR fluids and then introduces an interesting example of damping force controls utilizing a Preisach hysteresis compensator, followed by full vehicle suspension featuring an MR damper. Section 4.2 introduces geometric optimization of the MR damper, which considers advanced objective functions including damping force, dynamic range, and inductive time constant [12]. In recent years, research has been undertaken on MR damper design. Rosenfeld and Wereley [13] proposed an analytical optimization design method for MR valves and dampers based on the assumption of constant magnetic flux density throughout the magnetic circuit to ensure that one region of the magnetic circuit does not saturate prematurely and cause a bottleneck effect. Nevertheless, this assumption leads to a suboptimal result because valve performance depends on both the magnetic circuit and the geometry of the ducts through which the MR fluid passes.

Nguyen et al. [14] proposed an optimal finite element method (FEM)-based MR valve design constrained in a specified volume. This work considered the effects of all geometric variables of MR valves by minimizing the valve ratio calculated from the FE analysis. However, the control energy and time response of the MR valves was not considered in this research. Nguyen et al. [15] also studied an optimal MR valve design that satisfies specific operational requirements such as a pressure drop using minimum control energy. The inductive time constant was also taken into account as a state variable.

In this study, the power consumption of the valve was chosen as an objective function, while the pressure drop and inductive time constants, with their constraints, were treated as state variables. Therefore, in this section, it is necessary to convert the optimization problem using constrained state variables to an unconstrained one to increase computation time and use low-quality optimal design parameters. The cylindrical MR damper for vehicle suspension proposed by Lee and Choi [16] is considered in the test. After the configuration and working principle of the MR damper is considered, the damping force and dynamic range are derived from a quasi-static model based on the Bingham model of the MR fluid. The control energy and the inductive time constant are then obtained. The initial geometric dimensions of the damper are determined based on the assumption of constant magnetic flux density throughout the damper's magnetic circuit [13], and the objective function of the optimization problem is proposed based on the solution of the initial damper. The optimization procedure using a golden-section algorithm and a local quadratic fitting technique is subsequently constructed via ANSYS parametric design language (APDL). Using the developed optimization tool, optimal MR damper solutions, which are constrained in a specific cylindrical volume defined by its radius and height, are determined and presented.

Section 4.3 presents the hysteretic behavior of the MR damper due to the magnetic field and a new control strategy to achieve accurate damping force control performance [17]. As is well known, MR fluid inherently exhibits hysteretic nonlinear responses [18]. Many efforts have been made to account for the hysteretic behaviors of smart material actuators. One approach involves adopting robust control schemes to minimize the adverse effects of the hysteretic nonlinearity. In this case, model parameter variations caused by the hysteresis are normally treated as actuator uncertainties and are compensated by feedback control efforts. The other approach uses a nonlinear actuator-driving model to estimate the hysteretic effect that is compensated for in the feed-forward loop. The Preisach model is a potential candidate that can be employed for hysteresis modeling and applied to hysteresis compensation of MR fluids.

In this section, we consider hysteretic behavior that occurs due to applied magnetic fields (practically, input currents) in damping force control of MR dampers. For this purpose, a commercial MR damper, Delphi Corporation's Magneride™ (U.S.), is adopted and its damping force characteristics are experimentally evaluated. A Preisach hysteresis model for the MR damper is then established and its first order descending curves are experimentally identified. A feed-forward hysteretic compensator is subsequently formulated for damping force control and then experimentally verified. In addition, a quarter-vehicle suspension model is formulated and a sky-hook controller with the hysteretic compensator is implemented for the vibration control. Vibration control performances with and without a hysteretic compensator are experimentally evaluated in a quarter-vehicle test facility.

Section 4.4 discusses a full-vehicle suspension system featuring MR dampers [11]. To date, most analytical research into MR suspension systems have limited their scopes to a quarter-car model, while some experimental studies have focused on vibration isolation in half- or full-vehicle suspension systems. In practical situations in particular, robustness of MR suspension control systems robustness is easily subjected to parameter uncertainties and external disturbances. Consequently, the main contributions of this section involve constructing a mathematical model for a full-vehicle MR suspension system and demonstrating how the MR suspension system, susceptible to system uncertainties, can effectively attenuate the unwanted vibrations of a passenger vehicle. To accomplish this goal, a dynamic model for a full-vehicle suspension system with MR dampers is derived, and a controller is designed to reduce the vibration level caused by external road excitation. Control responses for vibration isolation are evaluated by adopting the HILS methodology and are presented in both time and frequency domains.

4.2 Optimal Design

4.2.1 Configuration and Modeling

The cylindrical MR damper for vehicle suspension proposed by Lee and Choi [16] is presented in Figure 4.1. The MR damper is divided into the upper and lower chambers by the piston and fully filled with the MR fluid. As the piston moves, the MR fluid flows from one chamber to the other through the orifices at both ends and the annular duct between the inner and outer cylinder. The gas chamber located outside acts as an accumulator of the MR fluid flow induced by the motion of the piston. By neglecting the frictional force between oil seals and assuming quasi-static behavior of the damper, the damping force of the MR damper can be expressed as:

$$F_d = P_2 A_p - P_1 (A_p - A_s) \tag{4.1}$$

where A_p and A_s are the piston and the piston-shaft areas, respectively. P_1 and P_2 are pressures in the upper and lower chamber of the damper, respectively. The relations between P_1, P_2, and the pressure in the gas chamber, P_a, can be expressed as:

$$P_2 = P_a + \Delta P_2; \quad P_1 = P_a - \Delta P_1 - \Delta P_3 \tag{4.2}$$

where ΔP_1, ΔP_2, and ΔP_3 are the pressure drops of MR flow through the lower MR valve orifice, the upper MR valve orifice, and the annular duct between

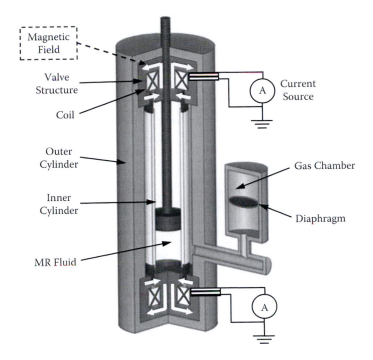

FIGURE 4.1
Configuration of the MR damper.

the outer and inner cylinder, respectively. The pressure in the gas chamber can be calculated as:

$$P_a = P_0 \left(\frac{V_0}{V_0 + A_s x_p} \right)^{\gamma}$$

(4.3)

where P_0 and V_0 are initial pressure and volume of the accumulator. γ is the coefficient of thermal expansion, which is ranging from 1.4 to 1.7 for adiabatic expansion. x_p is the piston displacement. By neglecting the minor loss, the pressure drops ΔP_1, ΔP_2, and ΔP_3 can be calculated as:

$$\Delta P_1 = \frac{12\eta L_m}{\pi t_m^3 R_1}(A_p - A_s)\dot{x}_p + 2c\frac{L_m}{t_m}\tau_y$$

$$\Delta P_2 = \frac{12\eta L_m}{\pi t_m^3 R_1}A_p\dot{x}_p + 2c\frac{L_m}{t_m}\tau_y$$

(4.4)

$$\Delta P_3 = \frac{6\eta L}{\pi t_g^3 R_2}(A_p - A_s)\dot{x}_p$$

where τ_y is the yield stress of the MR fluid induced by the applied magnetic field, η is the field-independent plastic viscosity (the base viscosity), L is the length of the inner cylinder, t_g is the gap of the annular duct between the inner and outer cylinder, R_1 and R_2 are the average radii of the intermediate pole and the annular duct, respectively, L_m is the length of the magnetic pole, t_m is the gap of the orifice of the MR valve structure, and c is the coefficient, which depends on flow velocity profile and has a value range from a minimum of 2.07 to a maximum of 3.07. The coefficient c can be approximately estimated as [19]:

$$c = 2.07 + \frac{12Q\eta}{12Q\eta + 0.8\pi R_1 t_m^2 \tau_y} \tag{4.5}$$

where Q is the flow rate of MR fluid flow through the valve structures. From Equation (4.1), Equation (4.2), and Equation (4.4), the damping force of the MR damper can be calculated by

$$F_d = P_a A_s + c_{vis}\dot{x}_p + F_{MR}\,\text{sgn}(\dot{x}_p) \tag{4.6}$$

where

$$c_{vis} = \frac{6\eta}{\pi}\left[\left(\frac{L}{R_2 t_g^3} + \frac{2L_m}{R_1 t_m^3}\right)(A_p - A_s)^2 + \frac{2L_m}{R_1 t_m^3}A_p^2\right]; \quad F_{MR} = (2A_p - A_s)\frac{2cL_m}{t_m}\tau_y$$

The first term in Equation (4.6) represents the elastic force from the gas compliance. The second term represents the damping force due to MR fluid viscosity. The third term is the force due to the yield stress of the MR fluid, which can be continuously controlled by the intensity of the magnetic field through the MR fluid ducts. This is the dominant term of the damping force, which is expected to have a large value in MR damper design. The dynamic range of the damper (defined as the ratio of the peak force with a maximum current input to the one with zero current input) can be approximately expressed as:

$$\lambda_d = \frac{c_{vis}\dot{x}_p + F_{MR}}{c_{vis}\dot{x}_p} \tag{4.7}$$

The dynamic range is also an important parameter in evaluating the overall performance of the MR damper. A large value of the dynamic range is expected to provide a wide control range of the MR damper.

In this test, the commercial MR fluid (MRF132-LD) from Lord Corporation is used. The induced yield stress of the MR fluid as a function of the applied magnetic field intensity (H_{mr}) is shown in Figure 4.2. By applying the least

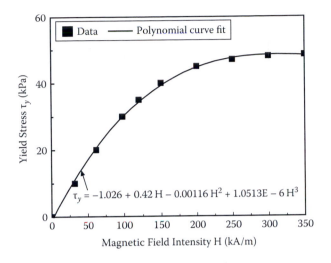

FIGURE 4.2
Yield stress of the MR fluid as a function of magnetic field intensity. (From Nguyen, Q.H. et al., *Smart Materials and Structures,* 18, 1, 2009. With permission.)

square curve fitting method, the yield stress of the MR fluid can be approximately expressed by

$$\tau_y = C_0 + C_1 H_{mr} + C_2 H_{mr}^2 + C_3 H_{mr}^3 \tag{4.8}$$

In Equation (4.8), the unit of the yield stress is kPa while that of the magnetic field intensity is KA/m. The coefficients C_0, C_1, C_2, and C_3 are identified as 0.3, 0.42, −0.00116, and 1.0513E−6, respectively. In order to calculate the damping force of the MR damper, it is necessary to solve the magnetic circuit of the damper. From the magnetic circuit solution, the yield stress of MR fluid in the active volume (the volume of the MR fluid where the magnetic flux crosses) can be obtained using Equation (4.8). The damping force is then calculated from Equation (4.6).

The magnetic circuit can be analyzed using the magnetic Kirchoff's law as:

$$\sum H_k l_k = N_c I \tag{4.9}$$

where H_k is the magnetic field intensity in the kth link of the circuit and l_k is the overall effective length of that link. N_c is the number of turns of the valve coil and I is the applied current in the coil wire. The magnetic flux conservation rule of the circuit is given by

$$\Phi = B_k A_k \tag{4.10}$$

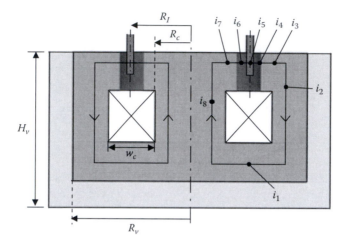

FIGURE 4.3
Approximate magnetic circuit of the MR damper.

where Φ is the magnetic flux of the circuit, and A_k and B_k are the cross-sectional area and magnetic flux density of the kth link, respectively. At low magnetic fields, the magnetic flux density, B_k, increases in proportion to the magnetic intensity H_k as: $B_k = \mu_0 \mu_k H_k$, where μ_0 is the magnetic permeability of free space ($\mu_0 = 4\pi 10^{-7} Tm/A$) and μ_k is the relative permeability of the kth link material. As the magnetic field becomes large, its ability to polarize the magnetic material diminishes and the material is almost magnetically saturated. Generally, a nonlinear B-H curve is used to express the magnetic property of material.

It is very difficult and complicated to find the exact solution of the magnetic circuit. Therefore, an approximate solution of the magnetic circuit is used in general. For the proposed MR damper, the approximate magnetic circuit is shown in Figure 4.3 and Equation (4.9) and Equation (4.10) can be rewritten as:

$$\sum_{i=1}^{8} H_i l_i = N_c I \tag{4.11}$$

$$\Phi = B_i A_i; \quad i = 1..8 \tag{4.12}$$

Due to small gap size of the MR ducts and small thickness of the intermediate pole, Equation (4.11) and Equation (4.12) can be approximately expressed as:

$$2H_1 l_1 + H_2 l_2 + H_{mr} l_{mr} + H_8 l_8 = N_c I \tag{4.13}$$

$$\Phi = B_1 A_1 = B_2 A_2 = B_{mr} A_{mr} = B_8 A_8 \tag{4.14}$$

H_{mr} and B_{mr} are effective magnetic field intensity and flux density of the MR fluid link, respectively. A_{mr} and l_{mr} are respectively the effective cross-sectional area and length of the MR link given by

$$l_{mr} = 2t_m; \quad A_{mr} = 2\pi R_1 L_m \tag{4.15}$$

From Equation (4.13) and Equation (4.14), the magnetic circuit of the valve can be solved based on the *B-H* curves of the MR fluid and the valve structure material. At low magnetic fields, the magnetic field intensity over the MR fluid link can be approximated as:

$$H_{mr} = \frac{N_c I}{2t_m + \dfrac{2l_1 \mu_{mr} A_{mr}}{\mu A_1} + \dfrac{l_2 \mu_{mr} A_{mr}}{\mu A_2} + \dfrac{l_8 \mu_{mr} A_{mr}}{\mu A_8}} \tag{4.16}$$

where μ_{mr} and μ are the relative permeability of MR fluid and the valve core material, respectively.

In order to improve the accuracy of the magnetic circuit solution, FEM has been employed [20, 21]. In this test, commercial FEM software, ANSYS, is used. Because the geometry of the valve structure is axisymmetric, a 2D-axisymmetric coupled element (PLANE13) is used for electromagnetic analysis. It is noted that a considerably accurate solution is expected by using FEM with a fine mesh because both a large number of links and the difference of flux density along the active MR volume are considered.

4.2.2 Design Optimization

In this section, a finite element method integrated with an optimization tool is used to obtain optimal geometric dimensions of the MR damper to minimize an objective function. For vehicle suspension design, the ride comfort and the suspension travel (the rattle space) are the two conflicting performance indexes to be considered. In order to reduce the suspension travel, high damping force is required. On the other hand, for improving ride comfort, low damping force is expected; thus, a large dynamic range is required. Furthermore, a fast time response MR damper is also expected in order to improve controllability of the suspension system. Considering these requirements, the following objective function is proposed:

$$OBJ = \alpha_F \frac{F_{MR,r}}{F_{MR}} + \alpha_d \frac{\lambda_{d,r}}{\lambda_d} + \alpha_T \frac{T}{T_r} \tag{4.17}$$

where F_{MR}, λ_d, and T are respectively the yield stress force, dynamic range, and inductive time constant of the damper, which are determined based on the FE solution of the damper magnetic circuit. $F_{MR,,r}$, $\lambda_{d,r}$, and T_r are the reference damping force, dynamic range, and inductive time constant of the damper, respectively. α_F, α_d, and α_T are respectively the weight factors for the damping force, dynamic range, and the inductive time constant ($\alpha_F + \alpha_d + \alpha_T = 1$). Of note, values of these weighting factors are chosen depending on each specific suspension system. For suspension systems designed for uneven or unpaved roads, large damping force is required. Thus, large values of α_F and α_T are chosen. On the other hand, a large value of α_d is used in the design of suspension systems for flat roads.

The inductive time constant and control energy of the MR damper can be calculated as:

$$T = L_{in}/R_w \tag{4.18}$$

$$N = I^2 R_w \tag{4.19}$$

where L_{in} is the inductance of the valve coil given by $L_{in} = N_c\Phi/I$, and R_w is the resistance of the coil wire that can be approximately calculated as:

$$R_w = L_w r_w = N_c \pi \bar{d}_c \frac{r}{A_w} \tag{4.20}$$

In Equation (4.20), L_w is the length of the coil wire, r_w is the resistance per unit length of the coil wire, $\bar{d}_c$ is the average diameter of the coil, A_w is the cross-sectional area of the coil wire, r is the resistivity of the coil wire, $r = 0.01726E\text{-}6\Omega m$ for copper wire, N_c is the number of coil turns, which can be approximated by $N_c = A_c/A_w$, and A_c is the cross-sectional area of the coil.

The reference damping force, dynamic range, and inductive time constant of the damper are obtained from the solution of the MR damper at initial values of design parameters. The initial geometric dimensions of the damper are determined based on the assumption of constant magnetic flux density throughout the magnetic circuit of the damper. Thus, the initial values of the core radius R_c, the coil width w_c, and the valve housing thickness t_h are determined as:

$$R_c = 2L_m; \quad w_c = \sqrt{R_v^2 - 4L_m^2} - 2L_m; \quad t_h = R_v - w_c - R_c \tag{4.21}$$

In Equation (4.21), R_v is the outside radius of the valve structure. The initial value of the pole length L_m is determined such that the magnetic flux density does not exceed the saturation values of the valve core material (silicon steel), which is 1.5 T in this test. It is noted that the permeability of the MR fluid is much smaller than that of the valve core material. Therefore, from Equation (4.16) the magnetic field intensity of the MR fluid link can be approximated by $H_{mr} = N_c I/2t_m$. In this case, in order to meet the saturation constraint of the valve core material, the following condition should be satisfied.

$$B_8 = B_{mr} \frac{A_{mr}}{A_8} = \frac{\mu_0 \mu_{mr} N_c I}{2t_m} \frac{A_{mr}}{A_8} \leq 1.5 \tag{4.22}$$

or

$$\frac{\mu_0 \mu_{mr} I \left(\sqrt{R^2 - 4L_m^2} - 2L_m \right)(H - 2L_m)}{2t_m A_w} \frac{R_1}{2L_m} \leq 1.5 \tag{4.23}$$

In order to find the optimal solution of the MR damper using FEM, firstly an analysis file for solving the magnetic circuit of the damper and calculating the objective function is built using ANSYS parametric design language (APDL). In the analysis file, the design variables (DVs) such as the coil width, the pole length, the MR orifice gap, and the core radius must be coded as variables and initial values are assigned to them. The geometric dimensions of the valve structure are varied during the optimization process, so that the meshing size is specified by the number of elements per line rather than element size. As aforementioned, the magnetic field intensity is not constant along the pole length. Therefore, it is necessary to define paths along the MR active volume where magnetic flux passes. The average magnetic field intensity across the MR ducts (H_{mr}) is calculated by integrating the field intensity along the defined path and then dividing by the path length. Thus, the magnetic flux and magnetic field intensity are determined as:

$$\Phi = 2\pi R_1 \int_0^{L_m} B_{mr}(s)\,ds; \quad H_{mr} = \frac{1}{L_m} \int_0^{L_m} H_{mr}(s)\,ds \tag{4.24}$$

where $B_{mr}(s)$ and $H_{mr}(s)$ are the magnetic flux density and magnetic field intensity at each nodal point on the defined path.

In the analysis, the first order method implemented in the ANSYS optimization tool is used to find the optimal solution. The procedures to achieve optimal design parameters of the MR damper using the first order method of ANSYS optimization tool are shown in Figure 4.4. Starting with initial value of DVs, by executing the analysis file, the magnetic flux, field intensity, damping force, dynamic range, inductive time constant, and the objective function are calculated. The ANSYS optimization tool then transforms the optimization problem with constrained design variables to unconstrained ones via penalty functions. The dimensionless, unconstrained objective function f is formulated as:

$$f(x) = \frac{OBJ}{OBJ_0} + \sum_{i=1}^{n} P_{x_i}(x_i) \tag{4.25}$$

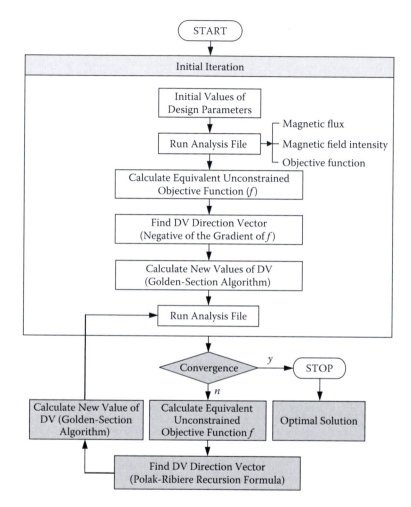

FIGURE 4.4
Flow chart to achieve optimal design parameters of the MR damper.

where OBJ_0 is the reference objective function value that is selected from the current group of design sets. P_{xi} is the exterior penalty function for the design variable x_i. For the initial iteration ($j = 0$), the search direction of DVs is assumed to be the negative of the gradient of the unconstrained objective function. Thus, the direction vector is calculated by

$$d^{(0)} = -\nabla f(x^{(0)}) \qquad (4.26)$$

The values of DVs in the next iteration ($j + 1$) is obtained from the following equation.

$$x^{(j+1)} = x^{(j)} + s_j d^{(j)} \qquad (4.27)$$

where the line search parameter s_j is calculated by using a combination of a golden-section algorithm and a local quadratic fitting technique. The analysis file is then executed with the new values of DVs and the convergence of the objective function is checked. If the convergence occurs, the values of the DVs at this iteration are optimum. If not, the subsequent iterations will be performed. In the subsequent iterations, the procedures are similar to those of the initial iteration except that the direction vectors are calculated according to the Polak-Ribiere recursion formula as:

$$d^{(j)} = -\nabla f(x^{(j)}) + r_{j-1} d^{(j-1)} \tag{4.28}$$

$$r_{j-1} = \frac{[\nabla f(x^{(j)}) - \nabla f(x^{(j-1)})]^T \nabla f(x^{(j)})}{|\nabla f(x^{(j-1)})|^2} \tag{4.29}$$

Thus, each iteration is composed of a number of sub-iterations that include search direction and gradient computations.

4.2.3 Optimization Results

In this section, the optimal solution for the MR damper is computed based on the optimization procedure in the previous section. Silicon steel, whose magnetic property is shown in Figure 4.5(a), is used for the valve core. The magnetic property of the MR fluid is shown in Figure 4.5(b). The coil wire is made of copper, whose relative permeability is assumed to be equal to that of free space, $\mu_c = 1$. The base viscosity of the MR fluid is assumed to be constant, $\eta = 0.092$ Pa.s, and the piston velocity used in the optimization problem is $v_p = 0.4$m/s. The significant dimensions of the MR damper are presented in Figure 4.6 in which the coil with w_c, the valve housing thickness t_h, the valve orifice gap t_m, and the pole length L_m are assigned as design variables. The coil wires are sized as 24-gauge (diameter = 0.5106 mm) and the maximum allowable current of the wire is 3 A. The maximum value of the applied current to the coil is $I = 2$ A. The current density applied to the coils can be approximately calculated by $J = I/A_w$.

As is well known, the smaller the mesh size is, the better the result is obtained. However, the smaller mesh size results in a high computational cost. When the mesh size is reduced to a certain value, the convergence of solution is expected. As aforementioned, the mesh size should be specified by the number of elements per line rather than element size. The number of elements on the lines across the MR fluid orifice is specified as a parameter called the basic meshing number. The number of elements of other lines is chosen as a product of the basic meshing number and an appropriate scalar. It is proved later that the basic meshing number of 10 is sufficient to ensure the convergence of the FE solution. Therefore, the basic meshing number of 10 is used in this test and the FE model of the valve structure at this

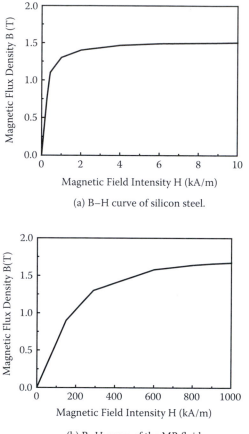

(a) B–H curve of silicon steel.

(b) B–H curve of the MR fluid

FIGURE 4.5
Magnetic properties of silicon steel and MR fluid. (From Nguyen, Q.H. et al., *Smart Materials and Structures,* 18, 1, 2009. With permission.)

basic meshing number is shown in Figure 4.7. Based on widely used MR dampers, the initial value of the MR orifice is chosen by $t_m = 0.75$ mm. From Equation (4.21) to Equation (4.23) and by assuming that the magnetic flux density of the silicon steel and MR fluid increases in proportion to the magnetic field intensity, $\mu = 1500$; $\mu_{mr} = 4$, the initial values of the design variables can be determined as: $L_m = 7.5$ mm; $w_c = 2.4$ mm; $t_h = 5.6$ mm. The magnetic flux density and magnetic flux obtained from the FEM solution are respectively shown in Figure 4.8(a) and (b). It is observed from the figure that the flux density of the magnetic circuit at the critical sections is approximately 1.43 to 1.52 T, which is about equal to the saturation flux density of the core material used in the previous analytical calculation (1.5 T). Thus, the results show a considerable agreement between the analytical and finite element solution.

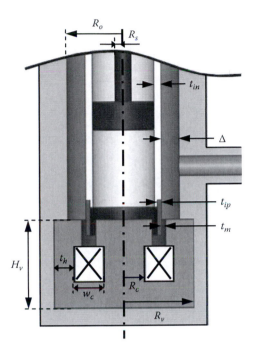

FIGURE 4.6
Geometric dimensions of the MR damper.

Figure 4.9 shows the optimal solution of the MR damper when the weighting factors are selected equally, $\alpha_F = \alpha_d = \alpha_T = 1/3$. The design variable limits are assigned as: 5 mm $\leq L_m \leq 13$ mm; 2 mm $\leq w_c \leq 8$ mm; 0.5 mm $\leq t_m \leq 1.5$ mm, and 3 mm $\leq t_h \leq 6$ mm. At these initial values, the yield stress force, dynamic range, inductive time constant, and objective function are 2065 N, 28.85, 50.5 ms, and 1, respectively. From the figure, it is observed that the solution is convergent after nine iterations at which the minimal value of the objective function is 0.943. At the optimum, the yield stress force, dynamic range, and inductive time constant are 1715 N, 35.64, and 40.6 ms, respectively. The optimal values of design variables L_m, w_c, t_m, and t_h in this case are 11.7 mm, 3 mm, 1.1 mm, and 4.05 mm, respectively. The results show that the dynamics range and conductive time constant are significantly improved. However, the yield stress force is a bit smaller than that at initial design variables. The power consumption of the damper is also calculated easily using Equation (4.19), which shows that the power consumption at the optimum (6.21 W) is smaller than that at the initial (8 W). Moreover, this power requirement can be achieved easily by a commercial power unit equipped for vehicles. Therefore, the power consumption was not included in the objective function.

As previously mentioned, for suspension systems designed especially for uneven or unpaved roads, large damping force is required and thus a large value of α_F should be used. Figure 4.10 shows the optimal solution of the MR

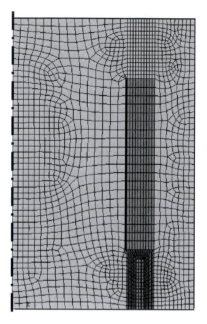

FIGURE 4.7
Finite element model for solving magnetic circuit of the damper. (From Nguyen, Q.H. et al., *Smart Materials and Structures*, 18, 1, 2009. With permission.)

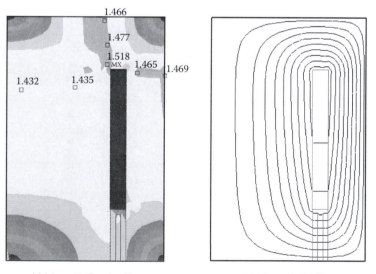

(a) Magnetic flux density   (b) Magnetic flux line

FIGURE 4.8
Finite element solution of the damper magnetic circuit at initial design variables. (From Nguyen, Q.H. et al., *Smart Materials and Structures*, 18, 1, 2009. With permission.)

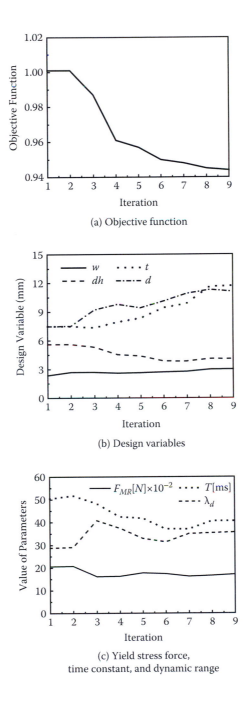

(a) Objective function

(b) Design variables

(c) Yield stress force,
time constant, and dynamic range

FIGURE 4.9
Optimized solution of the MR damper when the weighting factors are selected as follows: $\alpha F = \alpha d = \alpha T = 1/3$. (From Nguyen, Q.H. et al., *Smart Materials and Structures*, 18, 1, 2009. With permission.)

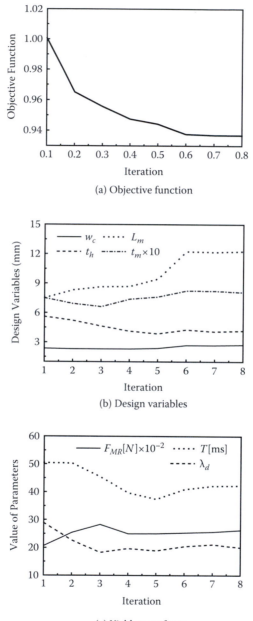

(a) Objective function

(b) Design variables

(c) Yield stress force,
time constant, and dynamic range

FIGURE 4.10

Optimized solution of the damper when the weighting factors are selected as follows: $\alpha F = 0.5$; $\alpha d = 0.2$; $\alpha T = 0.3$. (From Nguyen, Q.H. et al., *Smart Materials and Structures*, 18, 1, 2009. With permission.)

damper when the weighting factors are selected as: $\alpha_F = 0.5$; $\alpha_d = 0.2$; $\alpha_T = 0.3$. From the figure, it is observed that the solution is convergent after eight iterations. At the optimum, the damping force, dynamic range, inductive time constant, and objective function are 2630 N, 20, 42.3 ms, and 0.937, respectively. The optimal values of design variables L_m, w_c, t_m, and t_h in this case are 12.26 mm, 2.72 mm, 0.81 mm, and 4.17 mm, respectively. The results show that the yield stress force and conductive time constant are significantly improved while the dynamic range is smaller than that at initial design variables. Thus, by changing the weighting factors, a new optimal solution of the damper can be obtained in which the term with the larger value of the weighting factor will become dominant. The power consumption at the optimum is 5.16 W for this case.

In order to improve ride comfort, the damping force due to MR fluid viscosity (the damping force when no current is applied) should be small. Thus, a large value of dynamic range is required. Figure 4.11 shows the optimal solution of the MR damper when the weighting factors are selected as: $\alpha_F = 0.2$; $\alpha_d = 0.5$; $\alpha_T = 0.3$. The convergence of the solution occurs after 10 iterations. At the optimum, the damping force, dynamic range, power consumption, inductive time constant, and objective function are 1170 N, 63, 6.6 W, 40.3 ms, and 0.826, respectively. The optimal values of design variables L_m, w_c, t_m, and t_h in this case are 11.5 mm, 3.3 mm, 1.47 mm, and 4.1 mm, respectively. The results show that the purpose of improving dynamic range in order to improve ride comfort is well achieved in this case.

In order to prove the accuracy of the FE model, the dependence of the optimal solution on the meshing size is studied in this part. As is well known, the smaller the meshing size is, the better the result is obtained. However, the small meshing size results in a cost of computation time. When the meshing size is reduced to a certain value, a convergence of solution is expected. Figure 4.12 shows the dependence of the optimal solution on the meshing size (basic meshing number). It is observed that a convergence occurs when the basic meshing number is increased to 10 or more. An error of 0.15% is obtained when the basic meshing number is increased from 10 to 12. Thus, the results shown in Figure 4.9 to Figure 4.11 are considerably accurate.

For evaluating performance of the optimized MR dampers in the applied current range, the damping force and inductive time constant of initial and optimized dampers at different values of applied current are presented in Figure 4.13. In the figure, the terms "Opt.1", "Opt.2", and "Opt.3" represent the three optimized cases considered above, respectively: Opt.1 $\equiv \alpha_F = \alpha_d = \alpha_T = 1/3$; Opt.2 $\equiv \alpha_F = 0.5$, $\alpha_d = 0.2$, $\alpha_T = 0.3$; Opt.2 $\equiv \alpha_F = 0.2$; $\alpha_d = 0.5$; $\alpha_T = 0.3$. It is observed from the figure that the yield stress force approaches saturation when the applied current is increased above 1.5 A. Furthermore, at a small value of applied current, the inductive time constant is almost constant. The reason is that at small applied current, the induced magnetic flux density is small, so the magnetic flux density increases proportionally to the magnetic field intensity. At a high value of applied current, the magnetic flux

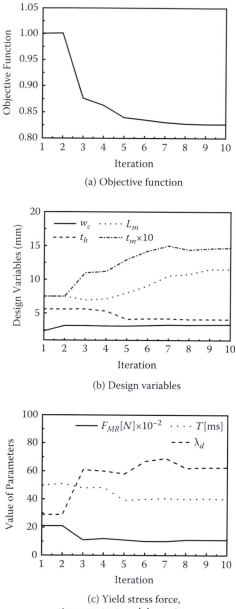

(a) Objective function

(b) Design variables

(c) Yield stress force,
time constant, and dynamic range

FIGURE 4.11
Optimized solution of the damper when the weighting factors are selected as follows: $\alpha F = 0.2$; $\alpha d = 0.5$; $\alpha T = 0.3$. (From Nguyen, Q.H. et al., *Smart Materials and Structures*, 18, 1, 2009. With permission.)

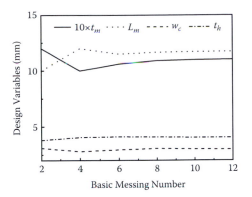

FIGURE 4.12
Dependence of the optimal solution on the meshing size. (From Nguyen, Q.H. et al., *Smart Materials and Structures*, 18, 1, 2009. With permission.)

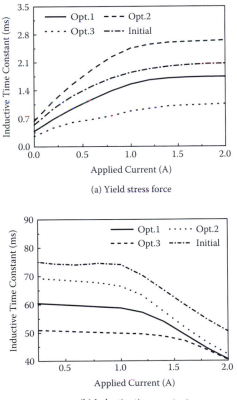

FIGURE 4.13
Yield stress force and inductive time constant of optimized and initial MR dampers at different values of applied. (From Nguyen, Q.H. et al., *Smart Materials and Structures*, 18, 1, 2009. With permission.)

density approaches the magnetic saturation of the material. This results in a saturation of yield stress force and a decrease of inductive time constant. It is noteworthy that the decrease of the inductive time constant is almost in proportion to the applied current.

4.3 Damping Force Control

4.3.1 MR Damper

A commercial MR damper, Delphi Corporation's Magneride™ that is used for high-class passenger vehicles, is adopted. Figure 4.14 shows the photograph of the MR damper featuring continuous controllability. Figure 4.15 presents the measured field-dependent damping force of the MR damper with respect to the piston velocity. This plot is obtained by calculating the maximum damping force at each velocity. The piston velocity is changed

FIGURE 4.14
Photograph of the commercial MR damper.

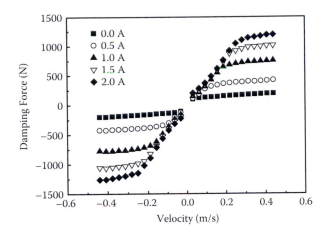

FIGURE 4.15
The field-dependent damping force of MR damper. (From Seong, M.S. et al., *Smart Materials and Structures,* 18, 7, 2009. With permission.)

by increasing the excitation frequency from 0.25 Hz to 3.5 Hz, while the excitation amplitude is maintained to be constant by ±20 mm. As expected, as the current (magnetic field) increases the damping force increases. We also observe that the MR damper has a damping force domain that covers the level of the damping force of a conventional hydraulic passive damper whose maximum value is around 1000 N. It is noted here that this damping force phenomenon is modeled using a biviscous model shown in the next section. Figure 4.16 presents the measured dynamic bandwidth of the

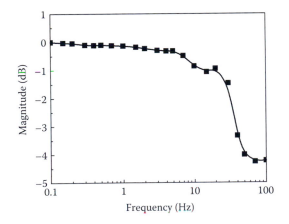

FIGURE 4.16
Dynamic bandwidth of MR damper. (From Seong, M.S. et al., *Smart Materials and Structures,* 18, 7, 2009. With permission.)

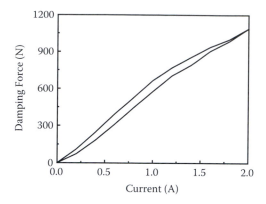

FIGURE 4.17

Hysteretic behavior of MR damper with respect to input current. (From Seong, M.S. et al., *Smart Materials and Structures*, 18, 7, 2009. With permission.)

damping force in the frequency domain. By sweeping the input current frequencies with the magnitude of 2 A, the dynamic bandwidth is obtained. It is identified that the dynamic bandwidth of the MR damper is approximately 38 Hz at –3 dB. Thus, both the vehicle body mode (1 to 2 Hz) and the wheel mode (10 to 15 Hz) of the passenger vehicle can be effectively controlled by employing the adopted MR damper.

The MR damper has two types of damping force hysteretic behavior. One is velocity-dependent hysteresis and the other is field-dependent hysteresis. Since a magnetic field (current) is directly connected with control input, the other largely affects control characteristics of the system. Figure 4.17 shows the measured hysteresis response of the MR damper with respect to input current. The direction of the hysteresis loop is counterclockwise. For purposes of this section, let us consider the Preisach model to identify hysteresis. The Preisach model was originally developed to represent hysteresis in magnetic materials, and is characterized by two significant properties—the minor loop property and the wiping-out property [22–24]. The minor loop property specifies that two comparable minor loops, which are generated by moving between two identical pairs of input maximum and minimum, are to be congruent if one exactly overlaps the other after some shift in the output parameter. The wiping-out property specifies which values of the preceding input trajectory affect the current output; that is, which dominant maximum and minimum can wipe out the effects of preceding smaller dominant maxima and minima. In order to determine whether the physical hysteresis phenomenon is matched by the Preisach model, one can examine the minor loop and wiping-out properties. These properties of MR fluid itself were evaluated by Han, Choi, and Wereley [18]. In the present work, we will evaluate the field-dependent hysteretic behavior of the damping force via the Preisach model and apply it to achieve accurate damping force control of the MR damper.

4.3.2 Preisach Model

The Preisach model that we adopt to describe the hysteretic behavior of an MR damper can be expressed as:

$$F_{MR}(t) = \iint_{\Gamma} \mu(\alpha, \beta)\gamma_{\alpha\beta}[i(t)]\, d\alpha\, d\beta \qquad (4.30)$$

where Γ is the Preisach plane, $\gamma_{\alpha\beta}[\cdot]$ is the hysteresis relay, $i(t)$ is the current (magnetic field), $F_{MR}(t)$ is the damping force due to magnetic field, and $\mu(\alpha, \beta)$ is the weighting function that describes the relative contribution of each relay to the overall hysteresis. Each relay is characterized by the pair of switching values (α, β) with $\alpha \geq \beta$. As the input varies with time, each individual relay adjusts its output according to the magnetic field input value, and the weighted sum of all relay outputs provides the overall system output. The simplest possible hysteretic relay for the MR damper is shown in Figure 4.18(a). It is a modification of a classical relay that has two

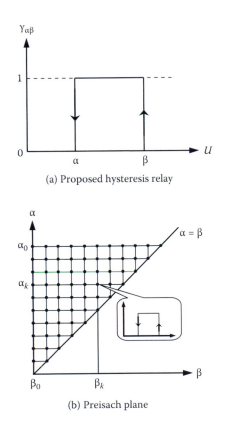

(a) Proposed hysteresis relay

(b) Preisach plane

FIGURE 4.18
Configuration of the Preisach model.

states, −1 and 1, corresponding to the opposite polarization of a ferromagnetic material. The output of the adopted relay is either 0 or 1. This is suitable for modeling hysteresis of an MR damper where the output (damping force) varies between zero and a maximum value. In this case, a hysteresis loop is located in the first quadrant of the input-output (current-damping force) plane. The Preisach plane can be geometrically interpreted as one-to-one mapping between relays and switching values of (α, β) as shown in Figure 4.18(b). The maximum damping force restricts the plane within the triangle in Figure 4.18(b), where α_0 and β_0 represent the upper and lower limit of the magnetic field input. In this work, they are chosen by 2A and 0A, respectively. The Preisach plane provides the state of individual relay, and thus the plane is divided into two time-varying regions as:

$$\Gamma_- = \{(\alpha, \beta) \in \Gamma \mid \text{output of } \gamma_{\alpha\beta} \text{ at } t \text{ is } 0$$
$$\Gamma_+ = \{(\alpha, \beta) \in \Gamma \mid \text{output of } \gamma_{\alpha\beta} \text{ at } t \text{ is } 1$$

(4.31)

The two regions represent that relays are on the 0 and 1 positions, respectively. Therefore, the first equation can be reduced to

$$F_{MR}(t) = \iint_{\Gamma} \mu(\alpha, \beta) \, d\alpha \, d\beta - \iint_{\Gamma_-} \mu(\alpha, \beta) \, d\alpha \, d\beta = \iint_{\Gamma_+} \mu(\alpha, \beta) \, d\alpha \, d\beta$$

(4.32)

The use of a numerical technique for the Preisach model identification has been normally proved as an effective way for smart materials [22, 24]. Therefore, in this work, the Preisach model for the MR damper is numerically implemented by employing a numerical function for each mesh value of (α, β) within the Preisach plane. The mesh values are identical with the data in the FOD curves, which are experimentally obtained. Figure 4.19(a) shows one of the mesh values of (α_1, β_1), and its corresponding FOD curve is a monotonic increase to a value α_1, then a monotonic decrease to β_1. After the input peaks at α_1, the decrease sweeps out area Ω, generating the descending branch inside the major loop. A numerical function $T(\alpha_1, \beta_1)$ is then defined as the output change along the descending branch as:

$$T(\alpha_1, \beta_1) = F_{MR}^{\alpha_1} - F_{MR}^{\alpha_1\beta_1}$$

(4.33)

From Equation (4.32) and Equation (4.33), we can determine an explicit formula for the hysteresis in terms of experimental data.

Figure 4.19(b) shows the increasing and decreasing series of an input magnetic field, where Γ_+ can be subdivided into n trapezoids Q_k. Geometrically, the area of each trapezoid Q_k can be expressed as the difference of two triangle

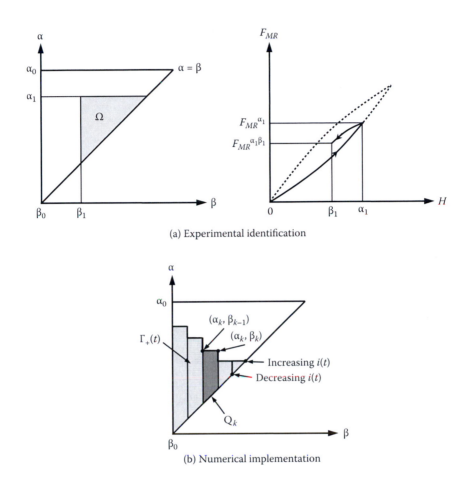

(a) Experimental identification

(b) Numerical implementation

FIGURE 4.19
Numerical identification and implementation of the Preisach model.

areas concerned with $T(\alpha_k, \beta_{k-1})$ and $T(\alpha_k, \beta_k)$, respectively. Therefore, the output for each Q_k is derived by

$$
\iint_{Q_k} \mu(\alpha, \beta)\, d\alpha\, d\beta = \iint_{\Omega_{k-1}} \mu(\alpha, \beta)\, d\alpha\, d\beta - \iint_{\Omega_k} \mu(\alpha, \beta)\, d\alpha\, d\beta
$$

$$
= \int_{\beta_{k-1}}^{\alpha_k} \int_{\beta}^{\alpha_k} \mu(\alpha, \beta)\, d\alpha\, d\beta - \int_{\beta_k}^{\alpha_k} \int_{\beta}^{\alpha_k} \mu(\alpha, \beta)\, d\alpha\, d\beta \qquad (4.34)
$$

$$
= T(\alpha_k, \beta_{k-1}) - T(\alpha_k, \beta_k)
$$

By summing the area of trapezoids Q_k over the entire area Γ_+, the output is expressed by

$$F_{MR}(t) = \sum_{k=1}^{n(t)} [T(\alpha_k, \beta_{k-1}) - T(\alpha_k, \beta_k)] \qquad \forall (\alpha_k, \beta_k) \in Q_k \qquad (4.35)$$

Consequently, for the case of the increasing and decreasing input, the output $F_{MR}(t)$ of the Preisach model is expressed by the experimentally defined $T(\alpha_k, \beta_k)$:

$$F_{MR,inc}(t) = \sum_{k=1}^{n(t)-1} [T(\alpha_k, \beta_{k-1}) - T(\alpha_k, \beta_k)] + T(i(t), \beta_{n(t)-1})$$

$$(4.36)$$

$$F_{MR,dec}(t) = \sum_{k=1}^{n(t)-1} [T(\alpha_k, \beta_{k-1}) - T(\alpha_k, \beta_k)] + [T(\alpha_{n(t)}, \beta_{n(t)-1}) - T(\alpha_{n(t)}, i(t))]$$

where the subscript *inc* means an increasing input and *dec* means a decreasing input. Numerical implementation of the Preisach model requires experimental determination of $T(\alpha_k, \beta_k)$ at a finite number of grid points within the Preisach plane. Practically, the finite number of grid points and the corresponding measured output values cause a certain problem that some values of the magnetic field input do not lie at the grid point. Furthermore, it may not have the measured output values. In this section, the following interpolation functions are employed to determine the corresponding $T(\alpha_k, \beta_k)$ when the value of a magnetic field input does not lie at the grid point:

$$T(\alpha_k, \beta_k)_{square\ cells} = c_0 + c_1 \alpha_k + c_2 \beta_k + c_3 \alpha_k \beta_k,$$

$$(4.37)$$

$$T(\alpha_k, \beta_k)_{triangular\ cells} = c_0 + c_1 \alpha_k + c_2 \beta_k$$

As a first step for hysteresis identification, FOD data sets are constructed to calculate the numerical function of the Preisach model. Figure 4.20(a) shows the input current (magnetic field) used to collect FOD curves. The current ranges from 0 A to 2 A, and is partitioned into 10 sub-ranges. The current is applied in a stepwise manner, and each step of the magnetic field is maintained for 5 s to satisfy a steady state condition. From the measured damping force using load cell, the FOD curves shown in Figure 4.20(b) are determined. At this point, the Preisach model for the MR damper is numerically identified using these FOD data sets.

Hysteresis prediction using the Preisach model is tested under two different types of input trajectories of the current: triangular and arbitrary signals.

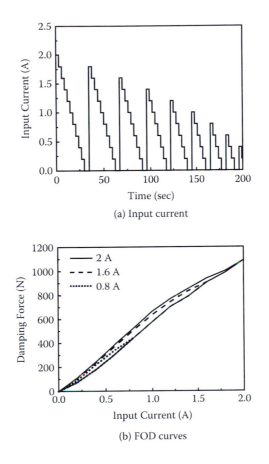

(a) Input current

(b) FOD curves

FIGURE 4.20
Measured FOD curves. (From Seong, M.S. et al., *Smart Materials and Structures*, 18, 7, 2009. With permission.)

Predictions using the Preisach model are compared to the results predicted by the Bingham model.

$$F_{MR}(i_B) = \alpha_B i_B^{\beta_B} \tag{4.38}$$

Bingham characteristics of the MR damper are experimentally measured and fitted to the function: $F_{MR}(i_B) = 486 i_B^{1.16}$.

Under the triangular input shown in Figure 4.21(a), the actual and predicted damping force responses are presented in Figure 4.21(b), whose magnifications are shown in Figure 4.21(c) and (d). In the case of increasing input, the predicted responses by the Preisach model and the Bingham model show similar results, but the hysteresis effect is distinctly manifested in the next decreasing step (refer to Figure 4.21(d)). It is observed from the results that the Preisach model well predicted the measured responses with great

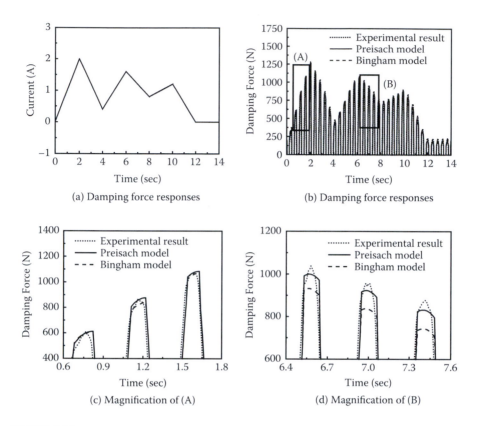

FIGURE 4.21
Actual and predicted damping force responses under triangular input. (From Seong, M.S. et al., *Smart Materials and Structures*, 18, 7, 2009. With permission.)

accuracy. Figure 4.22(b) presents the damping force behavior in response to the sinusoidal input trajectory in Figure 4.22(a). Its magnifications are shown in Figure 4.22(c) and (d). These results imply that the Preisach model accurately reconstructs the damping force hysteretic behavior of the MR damper.

4.3.3 Controller Formulation

In this section, a control algorithm is established to evaluate damping force control performance of the MR damper. Figure 4.23 shows the block diagram of the MR damper control system with the hysteretic compensator. At first, the desired damping force F_t is set in the system. The controllable damping force F_{MR} is determined using a biviscous model. In order to calculate the current i_B for the generation of controllable damping force, the inverse Bingham model is used. On the other hand, in order to account for the hysteresis, the feed-forward hysteretic compensator compensates the current i_P. Finally, control input i, which is determined by adding i_B and i_P to achieve

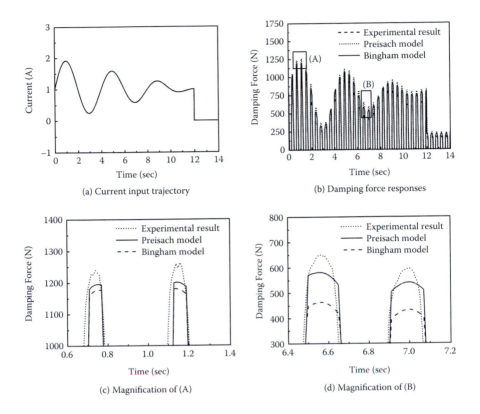

FIGURE 4.22
Actual and predicted damping force responses under sinusoidal input. (From Seong, M.S. et al., *Smart Materials and Structures*, 18, 7, 2009. With permission.)

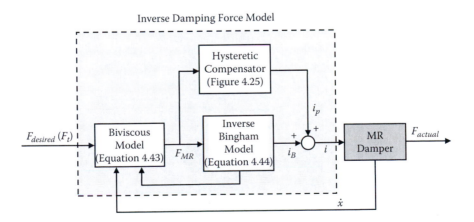

FIGURE 4.23
Block diagram for damping force control of MR damper.

the desired damping force, is applied to the MR damper to track the desired damping force. The implementation details of the control algorithm are described with three steps.

4.3.3.1 Biviscous model

The MR damper has a different damping coefficient between pre-yield stress velocity and post-yield stress velocity as shown in Figure 4.15. For representing this behavior, the biviscous model is adopted [25, 26]. A typical biviscous model is shown in Figure 4.24(a), where c_{pr} is the pre-yield stress damping coefficient, c_{po} is the post-yield stress damping coefficient, and $\dot{x}_y$ is the yield stress velocity. The yield stress velocity $\dot{x}_y$ depends on the input current; therefore, it can be expressed by

$$\dot{x}_y = a \cdot i_B^2 + b \cdot i_B + c \tag{4.39}$$

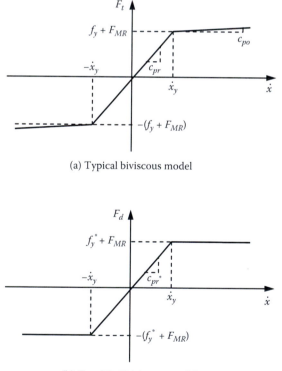

(a) Typical biviscous model

(b) Simplified biviscous model

FIGURE 4.24
Biviscous model.

where a, b, and c are obtained from the experiment as: $a = -0.0315$, $b = 0.18557$, and $c = 0.00773$. The total damping force of an MR damper F_t can be expressed using the biviscous model:

$$F_t = \begin{cases} c_{pr} \cdot \dot{x} & -\dot{x}_y \leq \dot{x} \leq \dot{x}_y \\ c_{po} \cdot \dot{x} + f_y + F_{MR} & \dot{x} > \dot{x}_y \\ c_{po} \cdot \dot{x} - (f_y + F_{MR}) & \dot{x} < -\dot{x}_y \end{cases} \qquad (4.40)$$

where c_{pr} and c_{po} are obtained from an experimental result: $c_{pr} = 3816$ Ns/m, $c_{po} = 190$ Ns/m. f_y can be calculated using the equation $f_y = c_{pr}\dot{x}_y$, $(i_B = 0)$. In order to simplify the damping force model Equation (4.40), we can divide the total damping force F_t as:

$$F_t = F_d + c_s\dot{x} \qquad (4.41)$$

Letting c_s be c_{po}, the simplified total damping force has no damping coefficient term after yield stress velocity as shown in Figure 4.24(b) and is mathematically represented as:

$$F_d = \begin{cases} c_{pr}^* \cdot \dot{x} & -\dot{x}_y \leq \dot{x} \leq \dot{x}_y \\ f_y^* + F_{MR} & \dot{x} > \dot{x}_y \\ -(f_y^* + F_{MR}) & \dot{x} < -\dot{x}_y \end{cases} \qquad (4.42)$$

where c_{pr}^* is $c_{pr} - c_{po}$ and f_y^* is $c_{pr}^*\dot{x}_y$, $(i_b = 0)$. From Equation (4.42), the damping force F_{MR} due to the magnetic field is obtained as:

$$F_{MR} = \begin{cases} 0 & -\dot{x}_y \leq \dot{x} \leq \dot{x}_y \\ F_d - f_y^* & \dot{x} > \dot{x}_y \\ -(F_d + f_y^*) & \dot{x} < -\dot{x}_y \end{cases} \qquad (4.43)$$

4.3.3.2 Inverse Bingham model

For calculating the current i_B, the inverse Bingham model is adopted using Equation (4.38) as:

$$i_B = (|F_{MR}|/\alpha_B)^{1/\beta_B} \qquad (4.44)$$

4.3.3.3 *Preisach hysteresis compensator*

The flow chart for implementation of a Preisach hysteretic compensator is shown in Figure 4.25. The proposed compensation algorithm consists of estimation and linearization of nonlinear hysteretic behavior using a Preisach hysteresis model. Detailed algorithms are listed as:

 I. Desired damping force of kth order $F_{MR,d}(k)$ is determined.
 II. Calculate current $i_B(k)$ using inverse Bingham model.

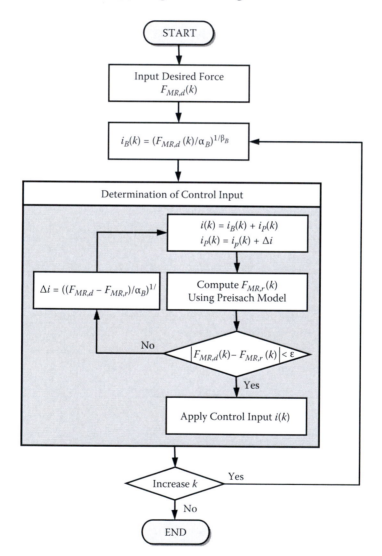

FIGURE 4.25
Flow chart for implementation of Preisach hysteretic compensator.

III. Determine the control current $i(k)$ by adding current $i_B(k)$ and $i_p(k)$.

IV. Determine actual damping force $F_{MR,r}(k)$ using a Preisach model and compare with desired damping force $F_{MR,d}(k)$.

V. Compare the difference of actual damping force $F_{MR,r}(k)$ and desired damping force $F_{MR,d}(k)$ with error limit ε.

VI. If the difference is larger than error limit ε, calculate the Δi and re-calculate control current $i(k)$.

VII. Repeat this algorithm while the difference is less than error limit ε.

VIII. When the difference is less than the error limit, apply the control current $i(k)$ to the MR damper.

In this work, the error limit ε is set to 5N, and to prevent the unlimited repeat, cycle limitation is set to 100.

4.3.4 Control Results

An experiment has been undertaken by realizing the control system shown in Figure 4.23 and Figure 4.25. Figure 4.26 shows the experimental configuration for damping force control of an MR damper. The MR damper is excited by amplitude ±20 mm and frequency 2.4 Hz using a hydraulic exciter. Damping force of the MR damper is measured by load-cell and piston movement is

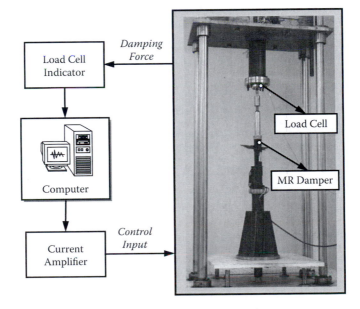

FIGURE 4.26
Experimental configurations for damping force control of MR damper.

measured by a linear variable differential transformer (LVDT). Control signal is generated from a computer data acquisition (DAQ) system, and this
signal is feedback to the current amp and applied to the MR damper.

Figure 4.27 presents the damping force control results under sinusoidal
desired damping force. Figure 4.27(b) shows the magnification graph of (A)
in Figure 4.27(a). It is seen from the control results that the proposed control
algorithm with hysteretic compensator gives more accurate damping force
control performance than without the hysteretic compensator. In addition,
control current with the hysteretic compensator is less than control current
without the compensator as shown in Figure 4.27(c). Figure 4.28 presents the
damping force control results under an arbitrary desired trajectory. Similar
control characteristics are observed in Figure 4.27. This accuracy in damping
force control can provide better vibration control performance in a suspension system. In the subsequent section, this will be investigated.

To investigate the effect of damping force controllability on the vibration
control of an MR suspension system, a quarter-vehicle suspension is adopted
as shown in Figure 4.29. It shows that a quarter-vehicle suspension model
with an MR damper has two degrees of freedom. The spring for the suspension is assumed linear and the tire is modeled as a linear spring component.
From the mechanical model, the state space equation is constructed for the
quarter-vehicle MR suspension system as:

$$\dot{\mathbf{x}} = \mathbf{A}\mathbf{x} + \mathbf{B}u + \mathbf{L}\dot{z}_r$$

$$\mathbf{y} = \mathbf{C}\mathbf{x}$$

(4.45)

where

$$x = [z_s - z_u \quad \dot{z}_s \quad z_u - z_r \quad \dot{z}_u]^T$$

$$A = \begin{bmatrix} 0 & 1 & 0 & -1 \\ -k_s/m_s & -c_s/m_s & 0 & c_s/m_s \\ 0 & 0 & 0 & 1 \\ k_s/m_u & c_s/m_u & -k_t/m_u & -c_s/m_u \end{bmatrix}$$

$$B = [0 \quad 1/m_s \quad 0 \quad -1/m_u]^T, C = [1 \quad 0 \quad 0 \quad 0], L = [0 \quad 0 \quad -1 \quad 0]^T$$

$$u = F_d$$

In the above, m_s and m_u are the sprung mass and unsprung mass, respectively. k_s is the stiffness coefficient of the suspension, c_s is the damping

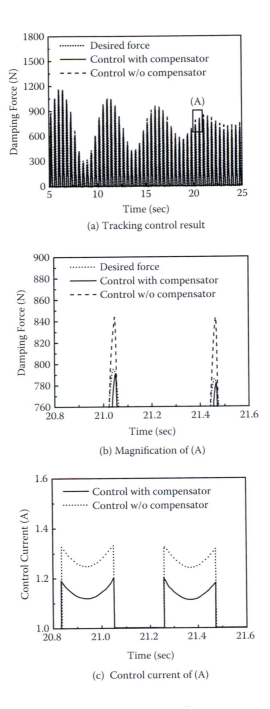

(a) Tracking control result

(b) Magnification of (A)

(c) Control current of (A)

FIGURE 4.27
Damping force control results for sinusoidal trajectory. (From Seong, M.S. et al., *Smart Materials and Structures*, 18, 7, 2009. With permission.)

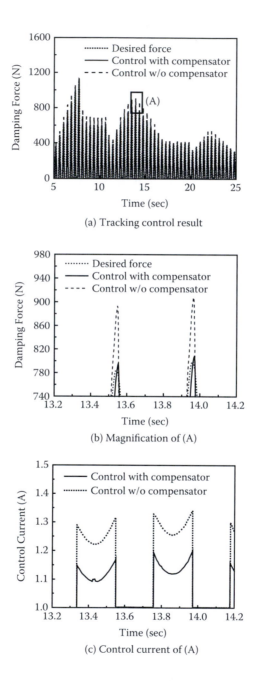

FIGURE 4.28
Damping force control results for arbitrary trajectory. (From Seong, M.S. et al., *Smart Materials and Structures*, 18, 7, 2009. With permission.)

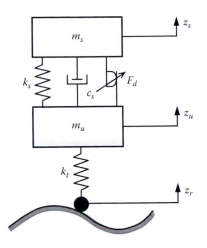

FIGURE 4.29
Mechanical model of the quarter-vehicle MR suspension system.

coefficient of the suspension, and k_t is the stiffness coefficient of the tire. z_s, z_u, and z_r are the vertical displacement of sprung mass, unsprung mass, and excitation, respectively.

In this test, a simple but very effective skyhook controller is adopted. It is well known that the logic of the skyhook controller is easy to implement to the real field. The desired damping force is set by

$$u = C_{sky}\dot{z}_s \qquad (4.46)$$

where C_{sky} is the control gain, which physically indicates the damping coefficient. In this work, this value is chosen by 7000 using a trial-and-error method. In Equation (4.45), u directly represents the damping force of F_d. On the other hand, the damping of the suspension system needs to be controlled depending upon the motion of suspension travel. Therefore, the following semi-active actuating condition is imposed.

$$u = \begin{cases} u & \text{for } u(\dot{z}_s - \dot{z}_u) > 0 \\ 0 & \text{for } u(\dot{z}_s - \dot{z}_u) \leq 0 \end{cases} \qquad (4.47)$$

Once the control input u is determined, the input current to be applied to the MR damper is obtained by damping force control algorithm.

Vibration control characteristics of the quarter-vehicle MR suspension system are evaluated under two types of excitation (road) conditions. The first excitation, normally used to reveal the transient response characteristic, is a bump described by

$$z_r = A_m[1 - \cos(\omega t)] \tag{4.48}$$

where $\omega = 2\pi f$, $f = 1/T$, $T = D/V$. $A_m (= 0.07 \text{ m})$ is the bump height, $D (= 0.8 \text{ m})$ is the width of the bump, and V is the vehicle velocity. In the bump excitation, the vehicle travels the bump with constant vehicle velocity of 3.08 km/h $(= 0.856 \text{ m/s})$. The second excitation is a sinusoidal function described by

$$z_r = A_m \sin \omega t \tag{4.49}$$

where ω (0.5 ~ 15 Hz) is the excitation frequency and A_m (1 ~ 10 mm) is the excitation amplitude.

Figure 4.30 shows an experimental apparatus of the quarter-vehicle MR suspension system to evaluate the effectiveness of vibration isolation of the control algorithm. The MR suspension (assembly of MR damper and spring), sprung mass, and tire are installed on the hydraulic system. The sprung mass displacement and suspension travel are measured by two LVDTs and excitation signal is measured by wire sensor. The hydraulic system applies the road profile to the MR suspension system. The current amplifier applies

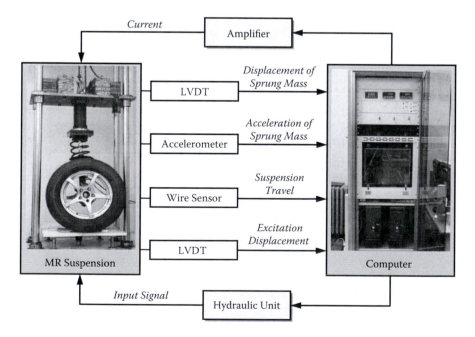

FIGURE 4.30
Experimental apparatus of the quarter-vehicle MR suspension system.

TABLE 4.1

System Parameters of the Quarter-Vehicle Suspension System

Parameter	Value
Sprung mass (m_s)	457.5 kg
Unsprung mass (m_u)	25.3 kg
Stiffness coefficient (k_s)	26,300 N/m
Damping coefficient (c_s)	190 Ns/m
Tire stiffness coefficient (k_t)	245,000 N/m

control current determined from control algorithm to the MR damper. The system parameters of the quarter-vehicle MR suspension system are chosen based on the design specifications of the conventional suspension system for a high-class passenger vehicle, and are listed in Table 4.1.

Figure 4.31(a) presents displacement versus time responses of the MR suspension system for the bump excitation and Figure 4.31(b) presents

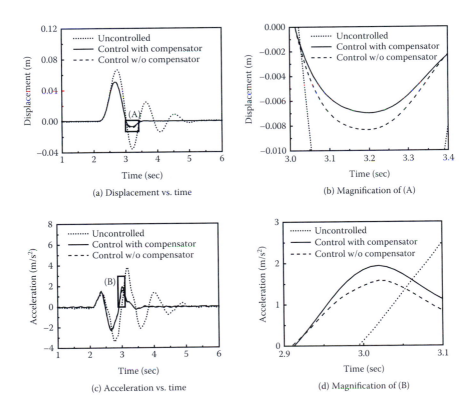

FIGURE 4.31

Control performances under bump excitation. (From Seong, M.S. et al., *Smart Materials and Structures*, 18, 7, 2009. With permission.) (*continued*)

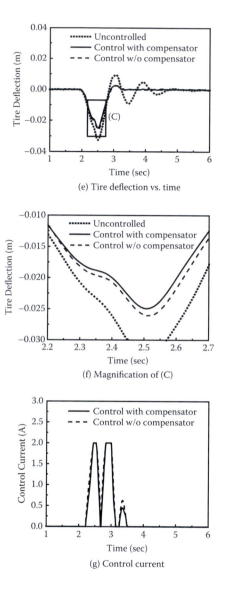

FIGURE 4.31 (*continued*)
Control performances under bump excitation. (From Seong, M.S. et al., *Smart Materials and Structures*, 18, 7, 2009. With permission.)

the magnification of (A). Figure 4.31(c) presents acceleration versus time responses and its magnification (specifically region "B") is presented in Figure 4.31(d). Figure 4.31(e) presents tire deflection versus time responses and its magnification (specifically region "C") is presented in Figure 4.31(f).

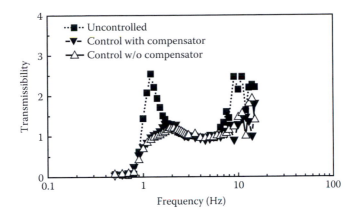

FIGURE 4.32

Transmissibility for suspension travel under sine excitation. (From Seong, M.S. et al., *Smart Materials and Structures*, 18, 7, 2009. With permission.)

As shown in the figures, it can obviously be found that unwanted vibrations induced from the bump excitation are well suppressed by adopting the Preisach hysteretic compensator to the MR suspension system. The input current with the hysteretic compensator is less than that without the compensator as shown in Figure 4.31(g). Figure 4.32 presents the transmissibility of the MR suspension system for the suspension travel ($z_s - z_u$). As expected, the transmissibility for the suspension travel has been substantially reduced in the neighborhood of the body resonance (1 ~ 2 Hz) by activating the controller. It is observed that vibration control performance can be further improved by adopting the hysteretic compensator.

4.4 Full-Vehicle Test

4.4.1 MR Damper

The schematic configuration of the MR damper and its photograph are shown in Figure 4.33. The MR damper is divided into the upper and lower chambers by the piston, and it is fully filled with the MR fluid. By the motion of the piston, the MR fluid flows through the orifice at both ends from one chamber to the other. The gas chamber, located outside, acts as an accumulator for absorbing sudden pressure variation of the lower chamber of the MR damper induced by the rapid motion of the piston. In order to simplify the

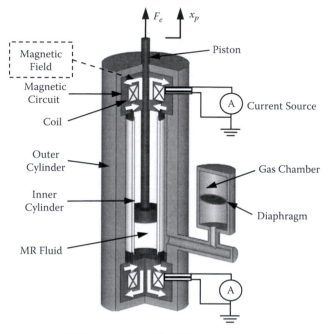

(a) Configuration of the MR damper

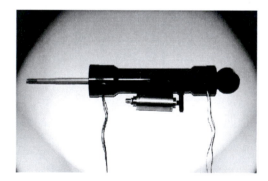

(b) Photograph of the MR damper

FIGURE 4.33
Configuration and photograph of the proposed MR damper.

analysis of the MR damper, it is assumed that the MR fluid is incompressible and that pressure in one chamber is uniformly distributed. Furthermore, it is assumed that frictional force between oil seals and fluid inertia are negligible. Thus, the damping force of the MR damper can be expressed as [27]:

$$F_e = k_e x_p + c_e \dot{x}_p + F_{MR} \operatorname{sgn}(\dot{x}_p) \qquad (4.50)$$

where

$$k_e = \frac{A_r^2}{C_g}, \quad c_e = (A_p - A_r)^2 \frac{12\eta L}{rh^3}, \quad F_{MR} = 4(A_p - A_r)\frac{2L_m}{h_m}\alpha H^\beta$$

In the above, x_p and $\dot{x}_p$ are the piston displacement and velocity, respectively. The magnetic field is related by $H = NI/2h_m$.

The first term in Equation (4.50) represents the spring force from the gas compliance, and the second damping force is due to the viscosity of the MR fluid. The third one is due to the yield stress of the MR fluid, which can be continuously controlled by the intensity of the magnetic field. In this test, for the MR fluid, the commercial product (MRF132-LD) from Lord Corporation is used and the experimental α and β are respectively identified as 0.083 and 1.25 by using a couette-type viscometer. Based on the Bingham model of the MR fluid, the size and the level of required damping force are determined so that the MR damper can be applicable to a middle-sized passenger vehicle. The principal design parameters of the MR damper are chosen as: the outer radius of the inner cylinder (r): 30.1 mm; the length of the magnetic pole (L_m): 10 mm; the gap between the magnetic poles (h_m): 1 mm; the number of coil turns (N): 150; and the diameter of the copper coil: 0.8 mm. Figure 4.34 presents the damping force with respect to the piston velocity at various magnetic fields. This plot is obtained by calculating the maximum damping force at each velocity. The detailed experimental apparatus and procedures to obtain this result have been well described in Reference [27]. It is clearly observed that the damping force is increased as the magnetic field increases. As a specific case, the damping force of 452 N at piston velocity of 0.38 m/s is increased up to 2189 N by applying the input current of 2.0 A. It is also seen that the simulated damping force agrees fairly well with the measured one. This advocates directly the validity of the MR damper model given by Equation (4.50). Figure 4.35 shows the measured dynamic response characteristic of the MR damper. It is obtained by applying constant current input (1.6 A) with various frequencies from 0.5 Hz to 100 Hz. It is identified that the dynamic response bandwidth of the MR damper is approximately 28 Hz at −3 dB. Thus, both the vehicle body mode (1 to 2 Hz) and the wheel

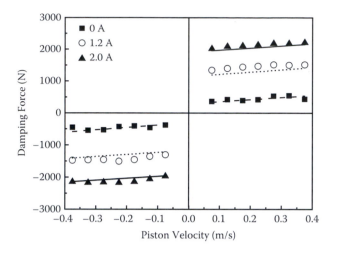

FIGURE 4.34
Comparison of the field-dependent damping force between the simulated and measured results. (From Choi, S.B. et al., *Vehicle System Dynamics*, 18, 7, 2002. With permission.)

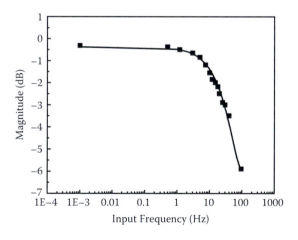

FIGURE 4.35
Dynamic response characteristic of the MR damper. (From Choi, S.B. et al., *Vehicle System Dynamics*, 18, 7, 2002. With permission.)

mode (10 to 13 Hz) of the passenger vehicle can be controlled effectively by employing the MR damper. Figure 4.36 presents power consumption of the MR damper required to generate a certain level of damping force. Both the damping force and required current are converted to mechanical and electrical energy, respectively. Approximately 8 W is required to generate the damping force of 2000 N. This amount of power can be supplied sufficiently from the commercial battery of passenger vehicles.

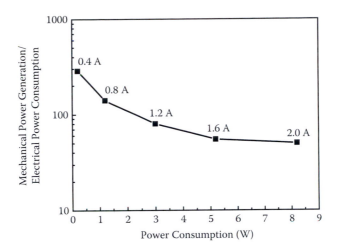

FIGURE 4.36
Mechanical power generation to electrical power consumption. (From Choi, S.B. et al., *Vehicle System Dynamics*, 18, 7, 2002. With permission.)

4.4.2 Full-Vehicle Suspension

We can construct a mathematical model for a full-vehicle MR suspension system equipped with four MR dampers as shown in Figure 4.37. The vehicle body itself is assumed rigid and has degrees of freedom in

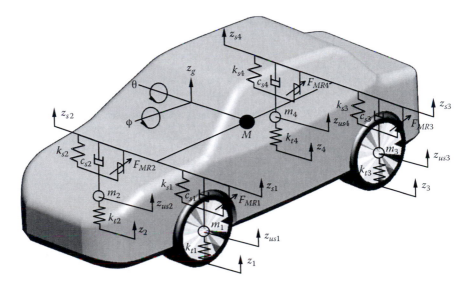

FIGURE 4.37
Mechanical model of the full-vehicle MR suspension system.

the vertical, pitch, and roll directions. It is connected to four rigid bodies representing the wheel unsprung masses in which each has a vertical degree of freedom. The bond graph method [28], which is very convenient for the modeling of hydraulic-mechanical systems, is used to obtain the governing equation of motion. From the bond graph model method, the governing equations of motion for the full-vehicle MR suspension are derived as:

$$M\ddot{z}_g = -f_{s1} - f_{s2} - f_{s3} - f_{s4} - F_{MR1} - F_{MR2} - F_{MR3} - F_{MR4}$$

$$J_\theta\ddot{\theta} = af_{s1} + af_{s2} - bf_{s3} - bf_{s4} + aF_{MR1} + aF_{MR2} - bF_{MR3} - bF_{MR4}$$

$$J_\phi\ddot{\phi} = -cf_{s1} + df_{s2} - cf_{s3} + df_{s4} - cF_{MR1} + dF_{MR2} - cF_{MR3} + dF_{MR4}$$

$$m_1\ddot{z}_{us1} = f_{s1} - f_{t1} + F_{MR1}$$

$$m_2\ddot{z}_{us2} = f_{s2} - f_{t2} + F_{MR2} \tag{4.51}$$

$$m_3\ddot{z}_{us3} = f_{s3} - f_{t3} + F_{MR3}$$

$$m_4\ddot{z}_{us4} = f_{s4} - f_{t4} + F_{MR4}$$

where $f_{si} = k_{si}(z_{si} - z_{usi}) + c_{si}(\dot{z}_{si} - \dot{z}_{usi})$ and $f_{ti} = k_{ti}(z_{usi} - z_i)$ for $i-1,2,3,4$. By defining the state as $x_{14\times1} = [z_g, \dot{z}_g, \theta, \dot{\theta}, \phi, \dot{\phi}, z_{us1}, \dot{z}_{us1}, z_{us2}, \dot{z}_{us2}, z_{us3}, \dot{z}_{us3}, z_{us4}, \dot{z}_{us4}]^T$, the control input vector as $u_{4\times1} = [F_{MR1}, F_{MR2}, F_{MR3}, F_{MR4}]^T$, and the disturbance vector as $w_{4\times1} = [z_1, z_2, z_3, z_4]^T$, we obtain the following state space equation:

$$\dot{x} = Ax + Bx + Lw$$

$$y = Cx \tag{4.52}$$

where $A \in \mathfrak{R}^{14\times14}$, $B \in \mathfrak{R}^{14\times4}$, $C \in \mathfrak{R}^{4\times14}$, and $L \in \mathfrak{R}^{14\times4}$ are the system matrix, the control input matrix, the output matrix, and the disturbance matrix, respectively. These matrices are given by

$$
A =
\begin{bmatrix}
0 & 1 & 0 & 0 & 0 & 0 & 0 & 0 & 0 & 0 & 0 & 0 & 0 & 0 \\[4pt]
S_{xg1} & S_{xgd1} & S_{th1} & S_{thd1} & S_{phi1} & S_{phid1} & 0 & 0 & 0 & 0 & 0 & 0 & 0 & 0 \\[4pt]
0 & 0 & 0 & 1 & 0 & 0 & \dfrac{ks_1}{M} & \dfrac{cs_1}{M} & \dfrac{ks_2}{M} & \dfrac{cs_2}{M} & \dfrac{ks_3}{M} & \dfrac{cs_3}{M} & \dfrac{ks_4}{M} & \dfrac{cs_4}{M} \\[8pt]
S_{xg2} & S_{xgd2} & S_{th2} & S_{thd2} & S_{phi2} & S_{phid2} & 0 & 0 & 0 & 0 & 0 & 0 & 0 & 0 \\[4pt]
0 & 0 & 0 & 0 & 0 & 1 & -a\dfrac{ks_1}{J_\theta} & -a\dfrac{cs_1}{J_\theta} & -a\dfrac{ks_2}{J_\theta} & -a\dfrac{cs_2}{J_\theta} & -a\dfrac{ks_3}{J_\theta} & -a\dfrac{cs_3}{J_\theta} & -a\dfrac{ks_4}{J_\theta} & -a\dfrac{cs_4}{J_\theta} \\[8pt]
S_{xg3} & S_{xgd3} & S_{th3} & S_{thd3} & S_{phi3} & S_{phi3} & 0 & 0 & 0 & 0 & 0 & 0 & 0 & 0 \\[4pt]
0 & 0 & 0 & 0 & 0 & 0 & c\dfrac{ks_1}{J_\phi} & c\dfrac{cs_1}{J_\phi} & -d\dfrac{ks_2}{J_\phi} & -d\dfrac{cs_2}{J_\phi} & c\dfrac{ks_3}{J_\phi} & c\dfrac{cs_3}{J_\phi} & -d\dfrac{ks_4}{J_\phi} & -d\dfrac{cs_4}{J_\phi} \\[8pt]
\dfrac{ks_1}{m_1} & \dfrac{cs_1}{m_1} & -a\dfrac{ks_1}{m_1} & -a\dfrac{cs_1}{m_1} & c\dfrac{ks_1}{m_1} & c\dfrac{cs_1}{m_1} & 0 & 1 & 0 & 0 & 0 & 0 & 0 & 0 \\[8pt]
0 & 0 & 0 & 0 & 0 & 0 & -\dfrac{ks_1+kt_1}{m_1} & -\dfrac{cs_1}{m_1} & 0 & 0 & 0 & 0 & 0 & 0 \\[8pt]
\dfrac{ks_2}{m_2} & \dfrac{cs_2}{m_2} & -a\dfrac{ks_2}{m_2} & -a\dfrac{cs_2}{m_2} & -d\dfrac{ks_2}{m_2} & -d\dfrac{cs_2}{m_2} & 0 & 0 & 0 & 1 & 0 & 0 & 0 & 0 \\[8pt]
0 & 0 & 0 & 0 & 0 & 0 & 0 & 0 & -\dfrac{ks_2+kt_2}{m_2} & -\dfrac{cs_2}{m_2} & 0 & 0 & 0 & 0 \\[8pt]
\dfrac{ks_3}{m_3} & \dfrac{cs_3}{m_3} & b\dfrac{ks_3}{m_3} & b\dfrac{cs_3}{m_3} & c\dfrac{ks_3}{m_3} & c\dfrac{cs_3}{m_3} & 0 & 0 & 0 & 0 & 0 & 1 & 0 & 0 \\[8pt]
0 & 0 & 0 & 0 & 0 & 0 & 0 & 0 & 0 & 0 & -\dfrac{ks_3+kt_3}{m_3} & -\dfrac{cs_3}{m_3} & 0 & 1 \\[8pt]
\dfrac{ks_4}{m_4} & \dfrac{cs_4}{m_4} & b\dfrac{ks_4}{m_4} & b\dfrac{cs_4}{m_4} & -d\dfrac{ks_4}{m_4} & -d\dfrac{cs_4}{m_4} & 0 & 0 & 0 & 0 & 0 & 0 & -\dfrac{ks_4+kt_4}{m_4} & -\dfrac{cs_4}{m_4}
\end{bmatrix}
$$

$$
B = \begin{bmatrix}
0 & 0 & 0 & 0 \\
-\dfrac{1}{M} & -\dfrac{1}{M} & \dfrac{1}{M} & \dfrac{1}{M} \\
0 & 0 & 0 & 0 \\
\dfrac{a}{J_\theta} & \dfrac{a}{J_\theta} & -\dfrac{b}{J_\theta} & -\dfrac{b}{J_\theta} \\
0 & 0 & 0 & 0 \\
\dfrac{c}{J_\phi} & -\dfrac{d}{J_\phi} & \dfrac{c}{J_\phi} & -\dfrac{d}{J_\phi} \\
0 & 0 & 0 & 0 \\
\dfrac{1}{m_1} & 0 & 0 & 0 \\
0 & 0 & 0 & 0 \\
0 & \dfrac{1}{m_2} & 0 & 0 \\
0 & 0 & 0 & 0 \\
0 & 0 & \dfrac{1}{m_3} & 0 \\
0 & 0 & 0 & 0 \\
0 & 0 & 0 & \dfrac{1}{m_4}
\end{bmatrix}
\qquad
L = \begin{bmatrix}
0 & 0 & 0 & 0 \\
0 & 0 & 0 & 0 \\
0 & 0 & 0 & 0 \\
0 & 0 & 0 & 0 \\
0 & 0 & 0 & 0 \\
0 & 0 & 0 & 0 \\
0 & 0 & 0 & 0 \\
\dfrac{k_{t1}}{m_1} & 0 & 0 & 0 \\
0 & 0 & 0 & 0 \\
0 & \dfrac{k_{t2}}{m_2} & 0 & 0 \\
0 & 0 & 0 & 0 \\
0 & 0 & \dfrac{k_{t3}}{m_3} & 0 \\
0 & 0 & 0 & 0 \\
0 & 0 & 0 & \dfrac{k_{t4}}{m_4}
\end{bmatrix}
$$

$$
C = \begin{bmatrix}
1 & 0 & -a & 0 & c & 0 & 0 & 0 & 0 & 0 & 0 & 0 & 0 & 0 \\
1 & 0 & -a & 0 & -d & 0 & 0 & 0 & 0 & 0 & 0 & 0 & 0 & 0 \\
1 & 0 & b & 0 & c & 0 & 0 & 0 & 0 & 0 & 0 & 0 & 0 & 0 \\
1 & 0 & b & 0 & -d & 0 & 0 & 0 & 0 & 0 & 0 & 0 & 0 & 0
\end{bmatrix}
$$

$$
S_{xg1} = \frac{-k_{s1} - k_{s2} - k_{s3} - k_{s4}}{M}
\qquad
S_{xgd1} = \frac{-c_{s1} - c_{s2} - c_{s3} - c_{s4}}{M}
$$

$$
S_{th1} = \frac{a \cdot k_{s1} + a \cdot k_{s2} - b \cdot k_{s3} - b \cdot k_{s4}}{M}
\qquad
S_{thd1} = \frac{a \cdot c_{s1} + a \cdot c_{s2} - b \cdot c_{s3} - b \cdot c_{s4}}{M}
$$

$$
S_{ph1} = \frac{-c \cdot k_{s1} + d \cdot k_{s2} - c \cdot k_{s3} + d \cdot k_{s4}}{M}
\qquad
S_{phd1} = \frac{-c \cdot c_{s1} + d \cdot c_{s2} - c \cdot c_{s3} + d \cdot c_{s4}}{M}
$$

$$
S_{xg2} = \frac{a \cdot k_{s1} + a \cdot k_{s2} - b \cdot k_{s3} - b \cdot k_{s4}}{J_\theta}
\qquad
S_{xgd2} = \frac{a \cdot c_{s1} + a \cdot c_{s2} - b \cdot c_{s3} - b \cdot c_{s4}}{J_\theta}
$$

$$S_{th2} = \frac{-a^2 \cdot k_{s1} - a^2 \cdot k_{s2} - b^2 \cdot k_{s3} - b^2 \cdot k_{s4}}{J_\theta} \quad S_{thd2} = \frac{-a^2 \cdot c_{s1} - a^2 \cdot c_{s2} - b^2 \cdot c_{s3} - b^2 \cdot c_{s4}}{J_\theta}$$

$$S_{ph2} = \frac{a \cdot c \cdot k_{s1} - a \cdot d \cdot k_{s2} - b \cdot c \cdot k_{s3} + b \cdot d \cdot k_{s4}}{J_\theta}$$

$$S_{phd2} = \frac{a \cdot c \cdot c_{s1} - a \cdot d \cdot c_{s2} - b \cdot c \cdot c_{s3} + b \cdot d \cdot c_{s4}}{J_\theta}$$

$$S_{xg3} = \frac{-c \cdot k_{s1} + d \cdot k_{s2} - c \cdot k_{s3} - d \cdot k_{s4}}{J_\varphi} \quad S_{xgd3} = -\frac{c \cdot c_{s1} + d \cdot c_{s2} - c \cdot c_{s3} - d \cdot c_{s4}}{J_\varphi}$$

$$S_{th3} = \frac{a \cdot c \cdot k_{s1} - a \cdot d \cdot k_{s2} - b \cdot c \cdot k_{s3} + b \cdot d \cdot k_{s4}}{J_\varphi}$$

$$S_{thd3} = \frac{a \cdot c \cdot c_{s1} - a \cdot d \cdot c_{s2} - b \cdot c \cdot c_{s3} + b \cdot d \cdot c_{s4}}{J_\varphi}$$

$$S_{ph2} = \frac{-c^2 \cdot k_{s1} - d^2 \cdot k_{s2} - c^2 \cdot k_{s3} - d^2 \cdot k_{s4}}{J_\varphi} \quad S_{phd2} = \frac{-c^2 \cdot c_{s1} - d^2 \cdot c_{s2} - c^2 \cdot c_{s3} - d^2 \cdot c_{s4}}{J_\varphi}$$

4.4.3 Controller Design

In practice, the sprung mass is varied by the loading conditions such as the number of riding persons and payload, and it makes the pitch and roll mass the moment of inertia change. In order to account for these structured system uncertainties, we adopt the loop shaping design procedure (LSDP) based on H_∞ robust stabilization proposed by McFarlane and Glover [29]. As a first step, we establish nominal and perturbed plants described by normalized left co-prime factorization as:

$$G = C(SI - A)^{-1}B = \tilde{M}^{-1}\tilde{N}$$

$$G = (\tilde{M} + \Delta_m)^{-1}(\tilde{N} + \Delta_n)$$

(4.53)

where $\tilde{M}$ and $\tilde{N}$ satisfy the following condition:

$$\tilde{M}\tilde{M}^* + \tilde{N}\tilde{N}^* = I$$

(4.54)

In the above, the symbol (*) denotes a complex conjugate transpose. In Equation (4.53), Δ_m and Δ_n satisfy the condition $\|[\Delta_m, \Delta_n]\|_\infty \le 1/\gamma$, where $1/\gamma$

TABLE 4.2

Parameters of a Full-Vehicle MR Suspension

Parameter	Value
M	1.33×10^3 kg
m_1, m_2	47 kg
m_3, m_4	40 kg
$c_{s1}, c_{s2}, c_{s3}, c_{s4}$	1.695×10^3 Nm/s
k_{s1}, k_{s2}	3.72×10^4 N/m
k_{s3}, k_{s4}	1.96×10^4 N/m
$k_{t1}, k_{t2}, k_{t3}, k_{t4}$	2×10^5 N/m
A_p	7.07×10^{-4} m^2
A_r	2.54×10^{-4} m^2
L_m	0.01 m
L	0.22 m
h_m	0.001 m
η	0.2 Pa·s

denotes the maximum allowable uncertainty bound. Based on the system parameters presented in Table 4.2, we impose the parameter variations of sprung mass and corresponding moment of inertia by 30%. Figure 4.38(a) presents the singular value of the nominal plant. It is seen that the magnitude is small in the low frequency, which reveals low gain and bandwidth. To reject effectively the disturbance and increase the closed-loop bandwidth, the following weighting function is designed:

$$W_c = \frac{1}{s+20} \tag{4.55}$$

We formulate now the shaped plant as:

$$G_{se} = GW$$

$$W = diag[1.5e6W_c, 1.5eW_c, 1.5eW_c, 1.5eW_c] \tag{4.56}$$

Figure 4.38(b) presents the singular value plot of the shaped plant. We see clearly that the magnitude is increased in the low frequency range, which results in robust performance to external disturbance.

As a second step, the optimal solution of γ_{min} for robust stabilization is computed for the shaped plant as:

$$\gamma_{min} = \left\{ 1 - \|\tilde{N}_{se}, \tilde{M}_{se}\|_H^2 \right\}^{-\frac{1}{2}} \tag{4.57}$$

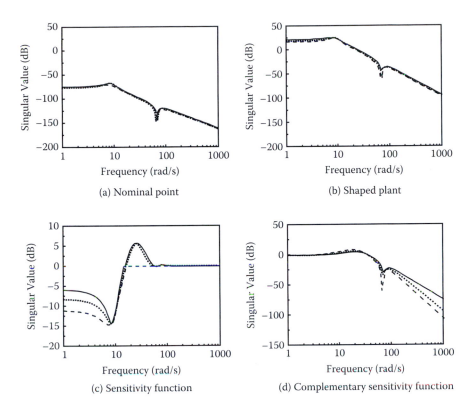

(a) Nominal point

(b) Shaped plant

(c) Sensitivity function

(d) Complementary sensitivity function

FIGURE 4.38
Singular value plots of the H_∞ controller. (From Choi, S.B. et al., *Vehicle System Dynamics*, 18, 7, 2002. With permission.)

where $\|\cdot\|_H$ denotes the Hankel norm. At this stage, γ can be viewed as a design indicator; if the loop shaping has been well carried out, a sufficient small value is obtained for γ_{min} [29]. By choosing a suitable γ little larger than γ_{min}, we obtain $\gamma = 3.22$ for the proposed system. Now, by using the value of γ, we can constitute the suboptimal controller K_∞ as [1]:

$$K_\infty = \left[\begin{array}{c|c} A_{se}^c + \gamma^2 W_h^{*-1} ZC_{se}^* (C_{se} + D_{se}F_h) & \gamma^2 W_h^{*-1} ZC_{se}^* \\ \hline B_{se}^* X & -D_{se}^* \end{array} \right] \quad (4.58)$$

$$A_{se}^c = A_{se} + B_{se}F_h, \quad W_h = I + (XZ - \gamma^2 I)$$

$$F_h = -S^{-1}(D_{se}^* C_{se} + B_{se}^* X), \quad S = I + D_{se}^* D_{se}$$

where A_{se}, B_{se}, C_{se}, and D_{se} are the state space system matrices of the shaped plant, G_{se}, X, and Z are the positive definite solutions of generalized control algebraic Riccati equation (GCARE) and generalized filtering algebraic

Riccati equation (GFARE), respectively. Finally, combining it with pre-designed weighting functions, the final H_∞ controllers are obtained as:

$$u = Ky$$
$$K = WK_\infty$$
(4.59)

Figure 4.38(c) and (d) present the sensitivity and complementary sensitivity functions. We get small magnitude in the low-frequency range and 0 dB in the high-frequency range for the sensitivity function plot. This result implies that the designed controller for each MR damper guarantees the desired performance such as effective suppression of unwanted vibration of the vehicle due to road input disturbances. Also from the complementary sensitivity function plot, we know that the sensor noise suppression and robustness to system uncertainties are well guaranteed because the plot shows small magnitude in the high-frequency range and 0 dB in the low-frequency range.

Figure 4.39 presents the block-diagram of the H_∞ control system. The control input determined from the H_∞ controller is to be applied to the MR damper depending upon the motion of suspension travel. Therefore, the following actuating condition is normally imposed:

$$u_i = \begin{bmatrix} u_i, \ for \ u_i(\dot{z}_{si} - \dot{z}_{usi}) > 0 \\ 0, \ for \ u_i(\dot{z}_{si} - \dot{z}_{usi}) \leq 0 \end{bmatrix} \quad (i = 1,2,3,4)$$
(4.60)

The above condition implies physically that the activating of the controller u_i only assures the increment of energy dissipation of the stable system [30].

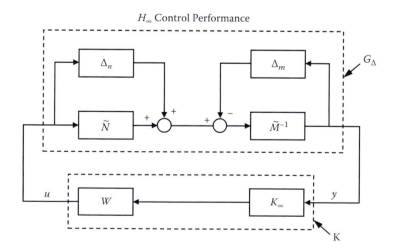

FIGURE 4.39
Block diagram of the LSDP H_∞ control system.

Once the control input u_i is determined, the input current to be applied to the MR damper is obtained by:

$$I_i = \frac{2h_m}{N}\left[u_i \cdot \frac{h_m}{2\alpha \cdot 4L_m(A_p - A_r)} \right]^{\frac{1}{\beta}} \quad (i = 1,2,3,4) \qquad (4.61)$$

4.4.4 Performance Evaluation

It generally takes a long time and high cost to develop successfully the new types of components for a complete system. To save money and time, a theoretical analysis using a computer simulation method is widely used. However, since many real situations that are difficult to be modeled and, even, cannot be modeled by the analytical method are often neglected or approximated by linearization, the theoretical method cannot precisely predict the performance of the system to occur in a real field. Therefore, in order to overcome the limit of the computer simulation method, the HILS were proposed recently. The HILS method has major advantages such as easy modification of system parameters and relatively low-cost test facilities. In addition, a wide range of operating conditions to emulate the practical situations can be investigated in the laboratory. In this test, an HILS is undertaken to evaluate the performance of the full-vehicle MR suspension system. The configuration of the proposed HILS is schematically shown in Figure 4.40.

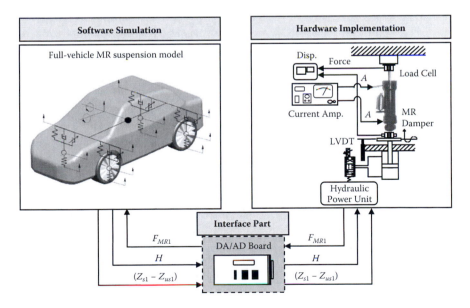

FIGURE 4.40
Schematic diagram of the hardware-in-the-loop-simulation (HILS) for the full vehicle suspension system.

The HILS is divided into three parts: interface, hardware, and software. The interface part is composed of a computer on which DSP board is mounted. The hardware part is composed of the MR damper, current amplifier, and hydraulic damper tester (hydraulic power unit). The software part consists of the theoretical model for the full-vehicle MR suspension system and control algorithm programmed in the computer. As a first step, the computer simulation for the full vehicle MR suspension system is performed with an initial value. This computer simulation is incorporated with the hydraulic damper tester, which applies the displacement (suspension travel) to the MR damper according to the demand signal obtained from the computer. It is also connected to the current amplifier, which applies control current determined from a control algorithm to the MR damper. Subsequently, the damping force of the MR damper is measured from the hydraulic damper tester and the measured damping force is fed back into the computer simulation. In short, the computer simultaneously runs both the hydraulic damper tester and the current amplifier during a simulation loop, and the computer simulation is performed based on the measured data. In this test, the MR damper at the front left side is chosen for the HILS by considering the capacity of the hydraulic damper tester. The measured damping force is fed back into the computer and simultaneously the damping forces of the other three MR dampers in the model are calculated by Equation (4.50) and Equation (4.51). In this calculation, it is assumed that the field-dependent dynamic characteristics of the three MR dampers are the same as one of the MR dampers used for hardware implementation.

Control characteristics for vibration suppression of the full-vehicle MR suspension system are evaluated under two types of road excitations through HILS. The first excitation normally used to reveal the transient response characteristic is a bump described by:

$$z_i = Z_b[1 - \cos(\omega_r t)] \quad (i = 1, 2)$$

$$z_j = Z_b[1 - \cos(\omega_r(t - D_{car}/V))] \quad (j = 3,4)$$

(4.62)

where $\omega_r = 2\pi V/D \cdot Z_b$ (= 0.035 m) is the half of the bump height, D (= 0.8 m) is the width of the bump, and D_{car} (= 2.4 m) is the wheel base, which is defined as the distance between the front wheel and the rear one. In the bump excitation, the vehicle travels the bump with a constant velocity of 3.08 km/h (= 0.856 m/s). The second type of road excitation, normally used to evaluate the frequency response, is a stationary random process with zero mean described by

$$\dot{z}_i + \rho_r V z_{Zi} = V W_{ni} \quad (i = 1,2,3,4)$$

(4.63)

where W_{ni} is white noise with intensity $2\sigma^2\rho_r V$. In random excitation, the values of road irregularity are chosen assuming that the vehicle travels on a paved road with the constant velocity of 72 km/h (= 20 m/s). The values of ρ_r = 0.45 m^{-1} and σ^2 = 300 mm^2 are chosen in the sense of the paved road condition. The system parameters of the MR suspension system are chosen based on the conventional suspension system for a middle-sized passenger vehicle, and are listed in Table 4.2.

Figure 4.41 presents time responses of the MR suspension system for the bump excitation. It is generally known that the displacement and acceleration of sprung mass and tire deflection are used to evaluate ride comfort

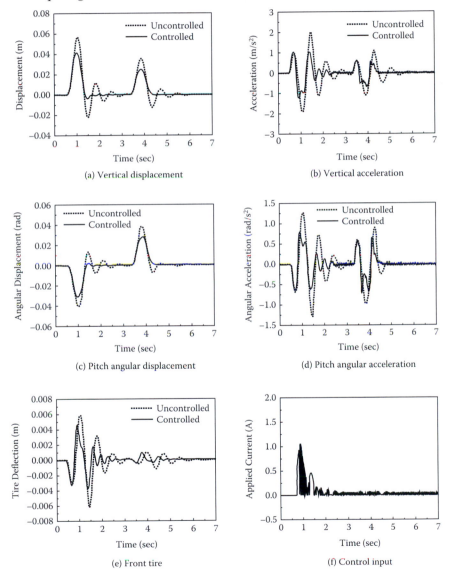

FIGURE 4.41

Bump responses of the MR suspension system via HILS. (From Choi, S.B. et al., *Vehicle System Dynamics*, 18, 7, 2002. With permission.)

and road holding of the vehicle, respectively. We see clearly that the vertical displacement and acceleration of sprung mass are reduced by employing the control current field determined from the H_∞ controller. In addition, we clearly observe that pitch angular displacement and acceleration and tire deflection are well reduced by applying the control input. Figure 4.42 presents controlled frequency responses using the H_∞ controller under random excitation. The frequency responses are obtained from power spectral density (PSD) for the vertical acceleration of the sprung mass and tire deflection. As expected, the power spectral densities for the vertical acceleration and tire deflection are respectively reduced approximately 20% and 40% in the neighborhood of body resonance (1 to 2 Hz) by applying input current as shown in Figure 4.42(c). In addition, it is seen that the vertical acceleration is also reduced approximately 30% at the wheel resonance (10 Hz). The control results presented in Figure 4.41 and Figure 4.42 indicate that both ride comfort and steering stability of a vehicle can be improved by employing the proposed semi-active MR suspension system.

4.5　Some Final Thoughts

In this chapter, we discussed a semi-active suspension system featuring MR shock absorbers (also called dampers) that can attenuate vehicle vibration from various road conditions to improve ride comfort as well as steering stability.

In Section 4.2, an optimal design procedure of MR shock absorber was established using ANSYS parametric design language. The optimization problem is to identify geometric dimensions of the valve structure employed in the MR damper to minimize an objective function. After obtaining the damping force and dynamic range of the damper based on the Bingham model of MR fluid, an objective function was proposed by a linear combination of the ratios of the damping force, dynamic range, and the inductive time constant to their reference values using the weighting factors. The reference values were determined based on the assumption of constant magnetic flux density throughout the magnetic circuit of the damper, which does not exceed the saturation magnetic flux density of the valve structure material. Optimal solutions of the damper with different value sets of the weighting factors depending on design purpose were obtained and presented. It was shown that by using different values of the weighting factors, various sets of optimal result can be obtained in which the term with the larger value of the weighting factor becomes dominant. The verification of the optimal solution convergence was also performed to validate the optimal results. In addition, the performance of the optimized MR dampers at different values of applied current was investigated showing the optimal trends of the yield stress force and inductive time constant.

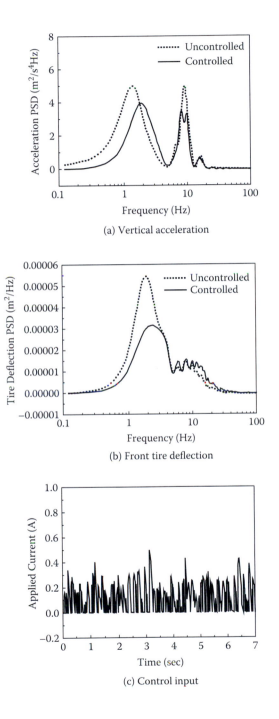

FIGURE 4.42
Random road responses of the MR suspension system via HILS. (From Choi, S.B. et al., *Vehicle System Dynamics*, 18, 7, 2002. With permission.)

Section 4.3 introduced a new control strategy for damping force control of MR damper and its effectiveness was experimentally verified. A feed-forward hysteretic compensator was formulated based on the Preisach hysteresis model of the MR damper. After evaluating damping force characteristics of a commercial MR damper (Delphi Magneride™), the control scheme was experimentally implemented. It has been demonstrated that the proposed control algorithm with the feed-forward hysteretic compensator produces more accurate damping force controllability as well as less power consumption. In addition, it has been shown through the quarter-car vehicle test that the MR suspension system with the skyhook controller integrated with the proposed hysteretic compensator can provide much better vibration control performance compared with the skyhook controller only. Hysteretic behavior of MR dampers due to the magnetic field is very important in accurate damping force control and thus should be considered. It is also remarked that in practice, other factors such as temperature variation or fluid sedimentation might have a similar overall effect on control performance. Therefore, these factors should be considered in damping force control of the MR damper.

A full-vehicle suspension system featuring MR dampers was proposed in Section 4.4. Its feedback control performance was evaluated via HILS. A cylindrical MR damper was designed and manufactured by incorporating the Bingham model of the MR fluid. After evaluating the field-dependent damping characteristics of the MR damper, a full-vehicle suspension system installed with four independent MR dampers was then constructed and its governing equations of motion were derived. A robust H_∞ controller was then designed to obtain a favorable control performance of the MR suspension system subjected to parameter uncertainties and external disturbances. Control characteristics for vibration suppression of the full-vehicle MR suspension system under various road conditions were evaluated through the HILS. For the bump excitation, the vibration levels represented by acceleration of the sprung mass and tire deflection were significantly reduced by implementing the H_∞ controller. For the random excitation, control characteristics were also remarkably enhanced by reducing the vertical acceleration of the sprung mass and tire deflection.

The results presented in this chapter are self-explanatory, justifying that the MR suspension system is very effective for vibration isolation of a passenger vehicle.

References

[1] Carlson, J. D., Cantanzarite, D. M., and St.Clair, K. A. 1996. Commercial magnetorheological fluid devices. *International Journal of Modern Physics B* 10: 2857–2865.

[2] Spencer Jr., B. F., Dyke, S. J., Sain, M. K., and Carlson, J. D. 1997. Phenomenological model for a magnetorheological damper. *Journal of Engineering Mechanics* 123: 230–238.

[3] Kamath, G. M., Wereley, N. M., and Jolly, M. R. 1998. Characterization of semi-active magnetorheological fluid lag mode damper. *SPIE Conference on Smart Structures and Integrated Systems*, pp.356–377.

[4] Yu, M., Liao, C. R., Chen, W. M., and Huang, S. L. 2006. Study on MR semi-active suspension system and its road testing. *Journal of Intelligent Material Systems and Structures* 17: 801–806.

[5] Guo, D. and Hu, H. 2005. Nonlinear stiffness of a magnetorheological damper. *Nonlinear Dynamics* 40: 241–249.

[6] Du, H., Sze, K. Y., and Lam, J. 2005. Semi-active H-infinity control of vehicle suspension with magnetorheological dampers. *Journal of Sound and Vibration* 238: 981–996.

[7] Shen, Y., Golnaraghi, M. F., and Heppler, G. R. 2007. Load-leveling suspension system with a magnetorheological damper. *Vehicle System Dynamics* 45: 297–312.

[8] Pranoto, T. and Nagaya, K. 2005. Development on 2DOF-type and rotary-type shock absorber damper using MRF and their efficiencies. *Journal of Materials Processing Technology* 161: 146–150.

[9] Ok, S. Y., Kim, D. S., Park, K. S., and Koh, H. M. 2007. Semi-active fuzzy control of cable-stayed bridges using magnetorheological dampers. *Engineering Structures* 29: 776–788.

[10] Choi, S. B., Lee, S. K., and Park, Y. P. 2001. A hysteresis model for the field-dependent damping force of a magnetorheological damper. *Journal of Sound and Vibration* 245: 375–383.

[11] Choi, S. B., Lee, H. S., and Park, Y. P. 2002. H-infinity control performance of a full-vehicle suspension featuring magnetorheological dampers. *Vehicle System Dynamics* 38: 341–360.

[12] Nguyen, Q. H. and Choi, S. B. 2009. Optimal design of a vehicle magneto-rheological damper considering the damping force and dynamic range. *Smart Materials and Structures* 18: 1–10.

[13] Rosenfield, N. C. and Wereley, N. M. 2004. Volume-constrained optimization of magnetorheological and electrorheological valves and dampers. *Smart Materials and Structures* 13: 1303–1313.

[14] Nguyen, Q. H., Han, Y. M., Choi, S. B., and Wereley, N. M. 2007. Geometry optimization of MR valves constrained in a specific volume using the finite element method. *Smart Materials and Structures* 16: 2242–2252.

[15] Nguyen, Q. H., Choi, S. B., and Wereley, N. M. 2008. Optimal design of magnetorheological valves via a finite element method considering control energy and a time constant. *Smart Materials and Structures* 17: 1–12.

[16] Lee, H. S. and Choi, S. B. 2000. Control and response characteristics of a magnetorheological fluid damper for passenger vehicles. *Journal of Intelligent Material Systems and Structures* 11: 80–87.

[17] Seong, M. S., Choi, S. B., and Han, Y. M. 2009. Damping force control of a vehicle MR damper using a Preisach hysteretic compensator. *Smart Materials and Structures* 18: 1–13.

[18] Han, Y. M., Choi, S. B., and Wereley, N. M. 2007. Hysteretic behavior of magnetorheological fluid and identification using Preisach model. *Journal of Intelligent Material Systems and Structures* 18: 973–981.

[19] Delivorias, P. P. 2004. Application of ER and MR fluid in an automotive crash energy absorber. Report No. MT04.18.

[20] Walid, H. E. A. 2002. Finite element analysis based modeling of magneto-rheological dampers. M.S. Thesis, Virginia Polytechnic Institute and State University, Blacksburg.

[21] Li, W. H., Du, H. and Guo, N. Q. 2003. Finite element analysis and simulation evaluation of a magnetorheological valve. *International Journal of Advances Manufacturing Technology* 21: 438–445.

[22] Mayergoyz, I. D. 1991. *Mathematical Models of Hysteresis*, New York: Springer-Verlag.

[23] Hughes, D. and When, J. T. 1994. Preisach modeling of piezoceramic and shape memory alloy hysteresis. *Smart Materials and Structures* 6: 287–300.

[24] Ge, P. and Jouaneh, M. 1997. Generalized Preisach model for hysteresis nonlinearity of piezoceramic actuators. *Precision Engineering* 20: 99–111.

[25] Wereley, N. M., Pang, L., and Kamath, G. M. 1998. Idealized hysteresis modeling of electrorheological and magnetorheological dampers. *Journal of Intelligent Material Systems and Structures* 9: 642–649.

[26] Ma, X. Q., Rakheja, S., and Su, C. Y. 2007. Development and relative assessments of models for characterizing the current dependent hysteresis properties of magnetorheological fluid dampers. *Journal of Intelligent Material Systems and Structures* 18: 487–502.

[27] Choi, S. B., Choi, Y. T., Chang, E. G., Han, S. J., and Kim, C. S. 1998. Control characteristics of a continuously variable ER damper. *Mechatronics* 8: 143–161.

[28] Karnopp, D. C., Margolis, D. L., and Rosenberg, R. C. 1991. *System Dynamics: A Unified Approach*, New York: John Wiley & Sons.

[29] McFarlane, D. C. and Glover, K. 1989. *Robust Controller Design Using Normalized Coprime Factor Plant Description*, Berlin: Springer-Verlag.

[30] Leitmann, G. 1994. Semiactive control for vibration attenuation. *Journal of Intelligent Material Systems and Structures* 5: 841–846.

5

Magnetorheological (MR) Suspension System for Tracked and Railway Vehicles

5.1 Introduction

Vehicle systems can be categorized into two types—wheeled vehicles and tracked vehicles. Wheeled vehicles include passenger vehicles and railway vehicles. As shown in Chapter 4, ride comfort and driving stability of passenger vehicles can be enhanced by adopting MR suspension systems, which attenuate vibrations from road conditions. We will now move on to suspension systems of other vehicles such as tracked vehicles.

A tracked vehicle is a vehicle that runs on continuous tracks instead of wheels. Tracked vehicles include construction vehicles, military armored vehicles, and unmanned ground vehicles. These vehicle systems must be operable on various terrain conditions, including soft terrain ground and off-road. High mobility tracked vehicle systems such as military tanks and armored vehicles must have good off-road mobility [1]. As the operating speed of high mobility tracked vehicle systems increases, the vibration induced by rough road conditions also increases. This induces fatigue on crewmembers and the many delicate instruments inside the vehicle. Vibration in a gun barrel reduces shooting accuracy.

Excessive vibration in high mobility tracked vehicles limits the maximum vehicle speed and consequently reduces survivability and operational efficiency in combat situations. Some studies on the vibration control of tracked vehicle systems with active and semi-active suspension systems are underway [2–4]. However, in view of MR suspension systems, most studies have focused on passenger vehicles. Research on the application of MR fluid to tracked vehicles is considerably rare. Consequently, the main contribution of Section 5.2 involves proposing a new suspension system for a tracked vehicle featuring an MR fluid-based valve [5]. In this section, a new double rod-type MR suspension unit (MRSU) is introduced and a two-coil annular MR valve is designed. Dynamic modeling of the proposed suspension system is undertaken and optimal design of an MR valve is performed to achieve principal valve design parameters. The optimization is undertaken using ANSYS and the solutions such as damping torque are presented. Moreover, vibration control responses of the tracked vehicle incorporated with MRSU are evaluated using computer simulation.

Section 5.3 introduces MR suspension systems for railway vehicles. The realization of high-speed railway vehicles, efficient transportation mechanisms for passengers and freight, is increasing in many countries. However, high-speed railway vehicles would cause car body vibration that may induce various problems such as ride stability, ride quality, and track abrasion. Thus, vibration control in railway vehicles is necessary for improving ride quality and car body stability. To overcome the performance limitations of passive suspension, many researchers have proposed the use of active suspension technology for oil valve and pneumatic actuators [1, 6–9].

A semi-active system utilizing an MR fluid damper was recently suggested as a potential candidate for railway vehicle suspensions. This section describes a feasibility study of the MR damper system as a secondary suspension system for railway vehicles. As a first step, an MR damper is designed and incorporated with the governing motion equation of the railway vehicle, which includes secondary suspension. Subsequently, via the skyhook control law, acceleration responses of the car body are analytically evaluated under track irregularities to illustrate the effectiveness of the secondary suspension used with the MR damper system in railway vehicles.

5.2 Tracked Vehicles

5.2.1 System Modeling

One of the representative passive suspension systems for the tracked vehicles is an in-arm suspension unit (ISU). The ISU has two types of mechanism configurations: single rod and double rod featuring greater load leveling capability and easy control of weight. In this section, the passive double rod ISU is modified in order to devise the MRSU. The schematic configuration of the proposed MRSU is shown in Figure 5.1. It is seen that an MR valve is attached to the orifice part of the passive ISU. Above a certain pressure difference between upper and lower chambers, the MR valve is activated by the magnetic field to achieve the desired damping forces [10].

In order to derive the piston velocity from external disturbance and moment arm of MRSU, the coordinate system of the MRSU is introduced in Figure 5.2. The coordinate system $x_i - y_i$ is fixed to point i and rotates with respect to the component. The coordinate (x_{ij}, y_{ij}) represents the point i with respect to the coordinate system $(x_j - y_j)$. Point 1 is the center of rotation of the MRSU, and point 2 is the center of the wheel. Point 6 represents the center of the actuating piston, and point 4 denotes the end of the connecting rod.

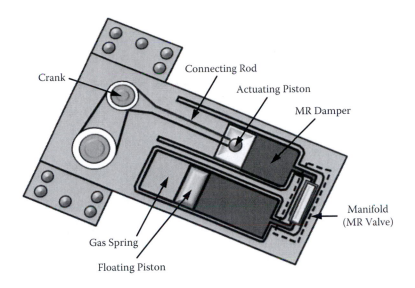

FIGURE 5.1
Configuration of the proposed MRSU.

Points 3 and 7 show the ends of the actuating cylinder. Point 5 is defined as the intersection of the perpendicular line from point 1 and the connecting rod. To get the position of the point, we can obtain x_{43}, y_{43} as follows:

$$x_{43} = x_{41} \cos(\theta_p) + y_{41} \sin(\theta_p)$$
$$y_{43} = l_v - x_{41} \sin(\theta_p) + y_{41} \cos(\theta_p) \tag{5.1}$$

where $x_{41} = l_{fx}$, $y_{41} = l_{fy}$, and $\theta_p = \theta + \theta_c$ as shown in Figure 5.2. The number 0 is assigned to y_{63} because point 6 is on the axis of the coordinate system $x_3 - y_3$. Therefore, we can derive the following equation:

$$y_{63} = y_{43} - l_d \sin(\theta_1) = 0 \tag{5.2}$$

Hence, $\theta_1 = \sin^{-1}(y_{43}/l_d)$. The piston position x_{63} with respect to the coordinate system $x_3 - y_3$ can be obtained as follows:

$$x_{63} = x_{43} - l_d \cos(\theta_1) \tag{5.3}$$

To drive the piston velocity, $\dot{x}_{43}$, $\dot{y}_{43}$ can be obtained as follows:

$$\dot{x}_{43} = (-x_{41} \sin(\theta_p) + y_{41} \cos(\theta_p))\dot{\alpha}$$
$$\dot{y}_{43} = (-x_{41} \cos(\theta_p) - y_{41} \sin(\theta_p))\dot{\alpha} \tag{5.4}$$

In addition, the angular speed $\dot{\theta}_1$ is defined by

$$\dot{\theta}_1 = \dot{y}_{43}/\sqrt{(l_d^2 - y_{43}^2)} \tag{5.5}$$

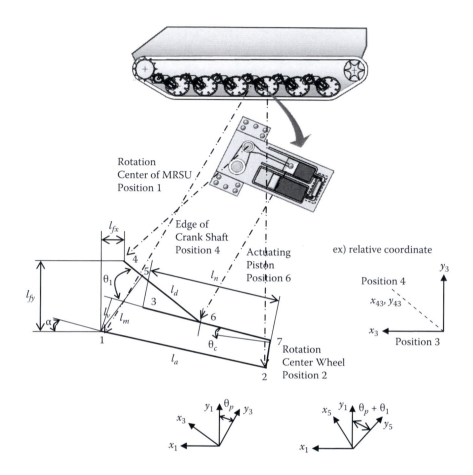

FIGURE 5.2
Coordinate system of an in-arm MRSU.

Thus, the piston velocity can be calculated by

$$\dot{x}_{63} = \dot{x}_{43} + l_d \sin(\theta_1)\dot{\theta}_1 \qquad (5.6)$$

The moment can be derived by the piston force and the distance between point 1 and point 5. The moment arm l_m is given by

$$l_m = -y_{15} = -[x_{41}\sin(\theta_p + \theta_1) - y_{41}\cos(\theta_p + \theta_1)] \qquad (5.7)$$

Therefore, the moment acting on point 1 can be obtained by

$$T_g = l_m F_s / \cos(\theta_1) \qquad (5.8)$$

where F_s is the spring force acting on the position.

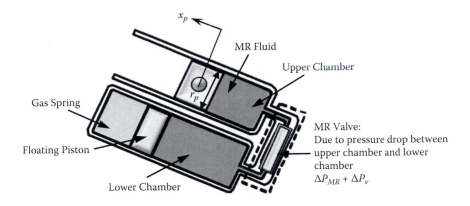

FIGURE 5.3
Dynamic motion diagram of MRSU.

To calculate the spring and damping torques of the proposed MRSU, consider the motion diagram as shown in Figure 5.3. By assuming that the gas does not usually exchange much heat with its surroundings, the gas pressure can be expressed by

$$P = P_{st}(V_{st}/V)^{1.4} \tag{5.9}$$

where P_{st} and $V_{st} = \pi r_g^2 l_g$ are the pressure and volume, respectively, in the static equilibrium state. The volume V of the gas chamber is given by

$$V = V_{st} + \pi r_p^2(x_p - x_{st}) \tag{5.10}$$

where x_p, x_{st} is the moving piston position and static piston position of each case. r_p is the radius of piston.

On the other hand, when the MR fluid flows through the MR valve, the pressure drop can be expressed by

$$\Delta P_m = \Delta P_{MR} + \Delta P_v \tag{5.11}$$

where ΔP_{MR} is the pressure drop due to the field-dependent yield stress of the MR fluid, and ΔP_v is the pressure drop due to the viscosity of the MR fluid. These are expressed by the following equations according to MR valve geometry.

$$\Delta P_{MR} = 2ct\tau_t/h + ca\tau_a/h \tag{5.12}$$

$$\Delta P_v = 6\eta LQ/\pi h^3 R_1 \tag{5.13}$$

where c is a coefficient that depends on the flow velocity profile, η is the base viscosity of MR fluid, Q is the flow rate through the MR valve, and R_1 is the average radius of the annular duct. Thus, the damping torque at the rotational center of the wheel arm is obtained as

$$\Delta P_m = \Delta P_{MR} + \Delta P_v \tag{5.14}$$

where F_d is the damping force at the actuating piston expressed by

$$F_d = \pi r_p^2 \Delta P_m \tag{5.15}$$

Now, the damping torque at the rotational center of the wheel arm is obtained as follows:

$$T_d = l_m F_d / \cos(\theta_1) \tag{5.16}$$

It should be remarked that in the derivation of the pressure equations, two assumptions are made: one is that the piston velocity is not very fast and the other is that the compressibility of the MR fluid can be ignored.

5.2.2 Optimal Design of the MR Valve

As mentioned, the passive double rod ISU is modified in order to devise the MRSU. It is seen from Figure 5.1 that an MR valve is attached to the manifold part of the passive ISU. The manifold of the MRSU consists of the MR valve and a two-coil annular valve. Figure 5.4 shows a simplified structure of the two-coil annular MR valve constrained specific volume. The valve geometry is characterized by the valve height L, the valve radius R, the valve housing thickness d_h, the MR channel gap h, the iron flange thickness t, and the

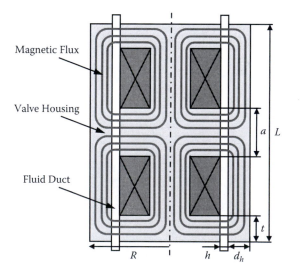

FIGURE 5.4
Magnetic circuit of the MR valve.

distance between the upper coil and lower coil *a*. When electric current is applied to the coil, a magnetic field appears as shown in Figure 5.4.

In order to optimize the MR valve, ANSYS integrated with an optimization tool is used to obtain optimal geometric dimensions of the MR valve of the MRSU using an objective function. The optimal objective is to minimize the valve ratio defined by the ratio of the viscous pressure drop to the field-dependent pressure drop of the MR valve. This ratio has a large effect on the characteristics of the MR valve. It is desirable that the valve ratio takes a small value. From Equation (5.12) and Equation (5.13), the valve ratios of the two-coil MR valve are calculated by

$$\lambda = \Delta P_v / \Delta P_{MR} = 3\eta LQ / \pi h^2 R_1 c(t\tau_t + 0.5a\tau_a) \tag{5.17}$$

The yield stress of the MR fluid caused by the magnetic circuit is measured and the polynomial curve of the yield stress is obtained as shown in Figure 5.5. In this work, a commercially available MR fluid (MRF-132DG manufactured by Lord Company) is used as shown in Table 5.1. The polynomial curve shown in Figure 5.5 can be expressed by the following equation:

$$\tau(kPa) = 52.962B^4 - 176.51B^3 + 158.79B^2 + 13.708B + 0.1442 \tag{5.18}$$

where τ is the yield stress caused by the applied magnetic field and B is the magnetic flux density of the applied magnetic field.

The geometry dimensions of the MR valves that affect the valve performance, such as the coil width, the flange thickness, and valve housing thickness, are considered as design variables (DV). At first, a log-file for solving the magnetic circuit of the valve and calculating the pressure drop and valve ratio using the ANSYS parametric design language is built. In this file, the design

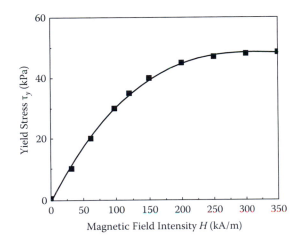

FIGURE 5.5
Yield stress of MR fluid.

TABLE 5.1

Magnetic Properties of the Valve Components

Valve Component	Material	Relative Permeability	Saturation Flux Density
Valve core	Silicon steel	2000	1.35 T
Valve housing	Silicon steel	1000	1.35 T
Coil	Copper	1	0
MR fluid	MRF-132DG	B-H Curve	B-H Curve
Nonmagnetic	Nonmagnetic	1	0
Cap/bobbin	Steel		

variables must be input as variables and initial values are assigned to them. The geometry dimensions of the valves vary during the optimization process, so the meshing size should be specified by the number of elements per line rather than element size. After solving for the magnetic circuit of the valve, the average magnetic flux density (B) through MR flows is calculated from the FE solution by integrating the flux density along a path and then dividing by the path length. The paths are defined along the ducts where the magnetic circuit passed. Figure 5.6 presents a flowchart of the optimization producers.

To achieve optimal design parameters of the MR valve, the first order method of the ANSYS optimization tool is used in this analysis. The ANSYS optimization tool transforms the constrained optimization problem into the unconstrained one via penalty functions. The dimensionless, unconstrained objective function is then formulated as follows [12]:

$$\lambda^{(*)}(y, p) = \lambda_1/\lambda_0 + \sum_{i=1}^{n} P_y(y_i) + q \sum_{i=1}^{m} P_g(g_i) \tag{5.19}$$

where λ_1 is the objective function, λ_0 is the reference objective function, q is the response surface parameter, P_x is the exterior penalty function, and P_g is the extended interior penalty function.

For the initial iteration, the search direction of design variable is assumed to be the negative of the gradient of the unconstrained objective function. The direction vector is calculated by

$$d^{(0)} = -\nabla\lambda^{(*)}(y^{(0)}, 1) \tag{5.20}$$

The values of design variables in the next iteration are calculated by line search parameter s_j

$$y^{(j+1)} = y^{(j)} + s_j d^{(j)} \tag{5.21}$$

where the line search parameter s_j is calculated using a combination of a golden-section algorithm and a local quadratic fitting technique. The log file

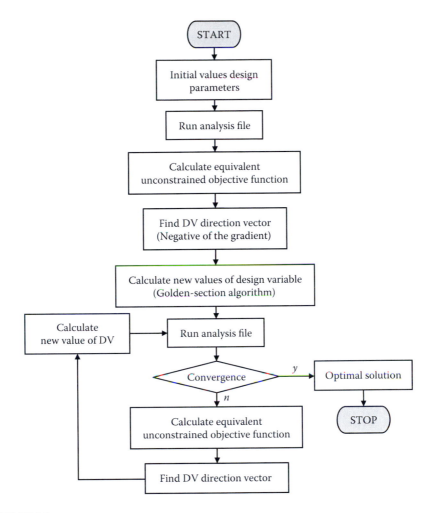

FIGURE 5.6
Flowchart to achieve optimal design parameter.

is executed with the new values of DVs and convergence of the objective function λ is checked. If convergence occurs, the values of DVs at the jth iteration are the optimum. If not, the subsequent iterations will be performed according to the following formula.

$$d^{(j)} = -\nabla\lambda^{(*)}(y^{(j)}, q_k) + u_{j-1}d^{(j-1)} \tag{5.22}$$

$$u_{j-1} = \{\nabla\lambda^{(*)}(y^{(j)}, q) - \nabla\lambda^{(*)}(y^{(j-1)}, q)\}^T \nabla\lambda^{(*)}(y^{(j)}, q)/\left|\nabla\lambda^{(*)}(y^{(j-1)}, q\right|^2 \tag{5.23}$$

Thus, each iteration is composed of a number of sub-iterations that include search direction and gradient computations.

Based on the optimization procedure in the previous equations, the optimal solution for an MR valve is achieved using the magnetic properties of an MR valve shown Table 5.1. The valves are constrained in a cylinder of radius $r = 23$ mm and height $L = 120$ mm, respectively. The base viscosity of the MR fluid is assumed to be constant, $\eta = 0.092$ Pas and the flow rate of the MR valves is $Q_0 = 3 \times 10^{-4} \, \text{m}^3 \text{s}^{-1}$. In this test, the valve gap is chosen by 1 mm. The current density applied to the coils can be calculated approximately by the following equation:

$$J = I/A_w \qquad (5.24)$$

where I is current applied to the coils and A_w is the cross-section of the coil wire. In this test, it is assumed the wire is sized as 24-gauge. The resistance per unit length of the wire is $0.18 \, \Omega\text{m}^{-1}$ and the maximum allowable current of the wire is 3 A. To evaluate power consumption of the valve coil, the power consumption N is expressed as follows:

$$N = I^2 R_w \qquad (5.25)$$

where R_w is the resistance of the coil wire, which can be calculated approximately as follows:

$$R_w = L_w r_w = V_c r_w / A_w \qquad (5.26)$$

where L_w is the length of the coil wire, r_w is the resistance per unit length of the coil wires, and V_c is the volume of all coils of the MR valve. The valve core radius, R_c, the iron flange thickness, t, and the valve housing thickness, d_h, are calculated by the following equations.

$$R_c = (1/2) \cdot \left\{ \sqrt{2R^2 - (W+h)^2} - (W+h) \right\} \qquad (5.27)$$

$$t = R_c / 2 \qquad (5.28)$$

$$d_h = R - (W + h + R_c) \qquad (5.29)$$

Figure 5.7 shows the optimal solution of a two-coil annular MR valve that is constrained in the specific volume when a current of 2.0 A is applied to the valve coils. The valve ratio, pressure drop, and power consumption at these initial values are $\lambda_0 = 0.087$, $\Delta P_{int} = 42.47$ bar, and $N_o = 24.51$ W according to coil width $W = 5$ mm, valve housing thickness $d = 4$ mm, and the iron flange thickness $t = 20$ mm, respectively. Optimal design variables are determined by $t_{opt} = 13.88$ mm, $W_{opt} = 1.5$ mm and $d_{h,opt} = 5.53$ mm. At these optimal design variables, the power consumption is $N_{opt} = 18.32$ W and the pressure drop is $\Delta P_{opt} = 67.23$ bar. Moreover, the difference between the magnetic flux density of the outer and inner duct is 0.0398 T, which is small and acceptable according to FEM analysis shown in Figure 5.8.

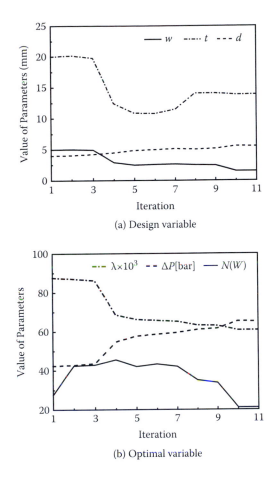

(a) Design variable

(b) Optimal variable

FIGURE 5.7
Optimization results of an MR valve.

5.2.3 Vibration Control Results

The vehicle model considered in this section consists of one axis of the military vehicle. It has a sprung mass and independent suspension. The vehicle body itself is assumed rigid and has degrees of freedom in vertical (X) directions. Figure 5.9 presents a mechanical model of a wheel and the body. The MRSU is modeled by the gas spring torque (T_g); MR damping torque (T_d) and the vertical force (F_y) are considered in the track. The governing equation of the model can be derived by

$$M_s \ddot{X}_s = -(T_g/l_m) - (T_d/l_m) + (k_s/\cos(\theta_1))X_u + (c_s/\cos(\theta_1))X_u$$

$$M_u \ddot{X}_u = (T_g/l_m) + (T_d/l_m) - (F_y/X)X_u - (k_s/\cos(\theta_1))X_u - (c_s/\cos(\theta_1))X_u + F_y$$

(5.30)

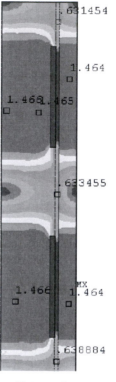

(a) Initial   (b) Optimal

FIGURE 5.8
Magnetic flux density of an MR valve.

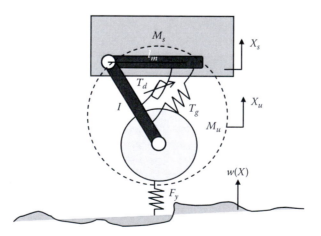

FIGURE 5.9
Mechanical model of a wheel and the body.

In Equation (5.30), M_s, M_u, F_y are the sprung mass, unsprung mass, and vertical force from the track, respectively. k_s, c_s are fundamental spring and damping coefficients through the initial pressure of the gas chamber and viscosity of the MR fluid.

The spring torque and damping torque of MRSU is compared between initial one and optimized one in Figure 5.10 and Figure 5.11. In the computer simulation, the parameters of MRSU are used as shown in Table 5.2. The spring torque is clearly seen in Figure 5.10 and Figure 5.11. The wheel torques represented by the spring torque exhibits nonlinear behavior in both cases. This is because the compressive effect of the gas chamber is proportional to the product of the piston area and the piston distance moved. It can also be seen that the spring characteristic depends upon the gas pressure in the chamber. In addition, the damping torque is increased as the

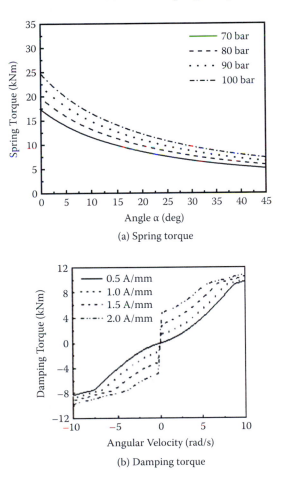

(a) Spring torque

(b) Damping torque

FIGURE 5.10
Characteristics of the MRSU: initial design.

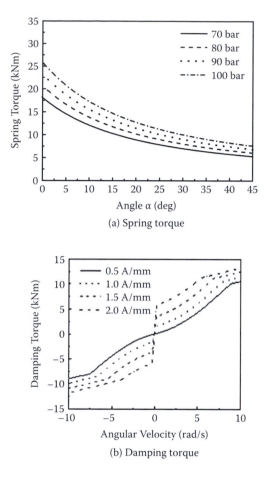

FIGURE 5.11
Characteristics of the MRSU: optimal design.

magnetic field increases. It is also observed that the damping torque of the optimized MRSU is larger than the initial one with the same control input. This means that the optimized MRSU has good performance in terms of energy efficiency.

Since the proposed MRSU is semi-active, it is easily expected that the performance of vibration isolation of the vehicle may be deteriorated if excessive magnetic fields are applied to the MRSU. Thus, it is necessary to use an appropriate control scheme to achieve satisfactory isolation performance under road conditions. Among many potential candidates for control algorithms, the semi-active skyhook controller is adopted in this work. It is well known that the logic of the skyhook is simple and easy

TABLE 5.2

Parameters of MRSU

Name	Symbol	Value
Height of valve	L	23 mm
Radius of valve	R	120 mm
Radius of actuating piston	r_p	76 mm
Duct thickness	h	1 mm
Flow flux of MR fluid	Q_0	$3 \times 10^{-4} m^3 s^{-1}$
Viscosity of MR fluid	η	0.092 Pas
Initial pressure of gas spring	P_{st}	1.01225e⁵Pa
Initial volume of gas spring	V_{st}	1.0053096e⁻³m³
Sprung mass	M_s	1.98 ton
Unsprung mass	M_u	128 kg
Vertical force	F_y	80,000 N

to implement in a practical field. The desired damping force for the MR damper is set by

$$u = C_{sky} \cdot X_s \qquad (5.31)$$

where C_{sky} is the gain of the skyhook controller. The control gain physically indicates the damping coefficient. On the other hand, the damping force of the suspension system needs to be controlled depending upon the motion of the suspension travel. Therefore, the following actuating condition is normally imposed.

$$u = \begin{bmatrix} C_{sky} \cdot X_s, & for\ \dot{X}_s(\dot{X}_s - \dot{X}_u) > 0 \\ 0, & for\ \dot{X}_s(\dot{X}_s - \dot{X}_u) \le 0 \end{bmatrix} \qquad (5.32)$$

For the computer simulation, the initial pressure of the gas spring is set to be 82 bars by considering the spring characteristics. Figure 5.12 presents bump responses of the tracked vehicle when the vehicle passes a 16-in. bump with a speed of 20 km/h. We clearly see that the vertical motion is substantially reduced by applying control current to the MRSU compared to the uncontrolled case (without control current). In addition, it is observed that the vibration reduction of the optimized MRSU is higher than the initial one although the control energy is smaller as shown in Figure 5.12(b). As shown in Figure 5.10(b) and Figure 5.11(b), the damping torque control range of the optimized MRSU is larger than the initial one with the same control input. This directly indicates that the optimized MRSU can provide better vibration isolation performance owing to the higher dynamic control range. From the results, it has now become evident to us that the MR suspension

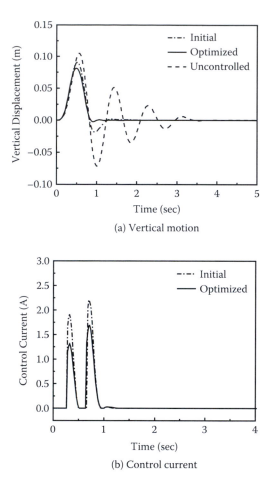

FIGURE 5.12
Bump response of the tracked vehicle.

system can be successfully installed not only to passenger vehicles but also to tracked vehicles to attenuate unwanted vibrations.

5.3 Railway Vehicles

5.3.1 System Modeling

The governing equations of motion for the railway vehicle with suspension systems can be derived using Newton's laws [11]. A 15-degree-of-freedom passenger railway vehicle model, shown in Figure 5.13, is presented to investigate the lateral response on a tangent track to random track irregularities.

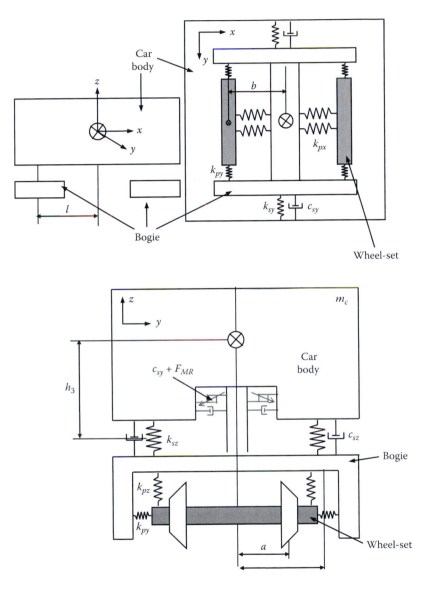

FIGURE 5.13
Mechanical model of a railway vehicle.

The model degrees of freedom are given in Table 5.3. The wheel-set of the railway vehicle is assumed to follow the track perfectly in the vertical direction; the wheel-set motion is given by the creep force input. A bogie frame roll is neglected as a degree of freedom. Wheel-rail interaction creep forces are calculated using Johnson and Vermeulen's creep theory. The vehicle equations of motion are presented, including the track input terms. The lateral car body response is then investigated for representative alignment

TABLE 5.3

The Degree of Freedom of a Railway
Vehicle

Element	Lateral	Yaw	Roll
1st Wheel-set	δ_1	δ_2	
2nd Wheel-set	δ_3	δ_4	
3rd Wheel-set	δ_7	δ_8	
4th Wheel-set	δ_9	δ_{10}	
1st Bogie frame	δ_5	δ_6	
2nd Bogie frame	δ_{11}	δ_{12}	
Car body	δ_{13}	δ_{14}	δ_{15}

and creep force input [8]. The governing equations for the wheel-set can be
expressed as follows:

$$m_w \ddot{\delta}_{1,3,7,9} + 2k_{py}(\delta_{1,3,7,9} - \delta_{5,12} - b\delta_{6,12}) + 2F_{y1,2,34} = 0$$

$$I_w \ddot{\delta}_{2,4,8,10} + 2k_{px}d_1^2(\delta_{2,4,8,10} - \delta_{6,12}) + 2F_{x1,2,3,4} = 0$$

(5.33)

The governing equations for the bogie frame can be expressed as follows:

$$m_f \ddot{\delta}_{5,11} - 2k_{py}(\delta_{1,7} - \delta_{5,11} - b\delta_{6,12}) - 2k_{py}(\delta_{3,9} - \delta_{5,11} + b\delta_{6,12})$$

$$+ 2k_{sy}(\delta_{5,11} - \delta_{13} - h_3\delta_{15} - l\delta_{14}) + 2c_{sy}(\dot{\delta}_{5,11} - \dot{\delta}_{13} - h_3\dot{\delta}_{15} - l\dot{\delta}_{14}) + 2F_{MR} = 0$$

$$I_{fy}\ddot{\delta}_{6,12} - b\{2k_{py}(\delta_{1,7} - \delta_{5,11} - b\delta_{6,12})\} + b\{2k_{py}(\delta_{3,9} - \delta_{5,11} + b\delta_{6,12})\}$$

$$- 2k_{px}d_1^2(\delta_{2,8} - \delta_{6,12}) - 2k_{px}d_1^2(\delta_{4,10} - \delta_{6,12}) = 0$$

(5.34)

The governing equations for the car body can be expressed as follows:

$$m_c \ddot{\delta}_{13} + 2k_{sy}(\delta_{13} - \delta_5 + h_3\delta_{15}) + 2c_{sy}(\dot{\delta}_{13} - \dot{\delta}_5 + h_3\dot{\delta}_{15}) + 2F_{MR}$$

$$+ 2k_{sy}(\delta_{13} - \delta_{11} + h_3\delta_{15}) + 2c_{sy}(\dot{\delta}_{13} - \dot{\delta}_{11} + h_3\dot{\delta}_{15}) + 2F_{MR} = 0$$

$$I_{cy}\ddot{\delta}_{14} + 2lk_{sy}(\delta_{13} - \delta_5 + h_3\delta_{15}) + 2lc_{sy}(\dot{\delta}_{13} - \dot{\delta}_5 + h_3\dot{\delta}_{15}) + 2lF_{MR}$$

$$- 2lk_{sy}(\delta_{13} - \delta_{11} + h_3\delta_{15}) - 2lc_{sy}(\dot{\delta}_{13} - \dot{\delta}_{11} + h_3\dot{\delta}_{15}) - 2lF_{MR} = 0$$

$$I_{cr}\ddot{\delta}_{15} + 2h_3 k_{sy}(\delta_{13} - \delta_5 + h_3\delta_{15}) + 2h_3 c_{sy}(\dot{\delta}_{13} - \dot{\delta}_5 + h_3\dot{\delta}_{15}) + 2h_3 F_{MR}$$

$$+ 2h_3 k_{sy}(\delta_{13} - \delta_{11} + h_3\delta_{15}) + 2h_3 c_{sy}(\dot{\delta}_{13} - \dot{\delta}_{11} + h_3\dot{\delta}_{15}) + 2h_3 F_{MR} = 0$$

(5.35)

where $m_{c,f,w}$ and $I_{c,f,w}$ are the mass and inertia moment of car body, bogie frame,
and wheel-set, respectively. $K_{p,s}$ and c_s are the stiffness and damping ratio of

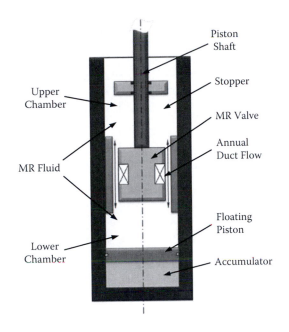

FIGURE 5.14
Configuration of an MR damper.

primary and secondary suspension. l, h are distance between bogie frame and car body. The MR damping force and creep force are expressed by F_{MR} and $F_{x,y}$.

The configuration of the proposed MR suspension system is shown in Figure 5.14. The MR suspension system is divided into damper parts and spring parts. The damper parts are divided into lower chamber and upper chamber through the piston head. The MR damper and safety damper are fully filled with MR fluid. By the motion of the piston, the MR fluid flows through the orifice at both ends from upper chamber to lower chamber. In order to simplify the analysis of the MR damper, it is assumed that the MR fluid is incompressible and that pressure in one chamber is uniformly distributed. Furthermore, it is assumed that frictional force between oil seals and fluid inertia are negligible. By assuming quasi-static behavior of the damper, the damping force can be expressed as follows:

$$F_d = P_2 A_p - P_1 (A_p - A_s) \tag{5.36}$$

where A_p and A_s are the piston and the piston shaft effective cross-sectional areas, respectively. P_1 and P_2 are pressures in the upper and lower chamber of the MR damper, respectively. The relations P_1 (or P_2) with P_a, the pressure in the accumulator, can be expressed as follows:

$$P_2 = P_a$$
$$P_1 = P_a - \Delta P_v \tag{5.37}$$

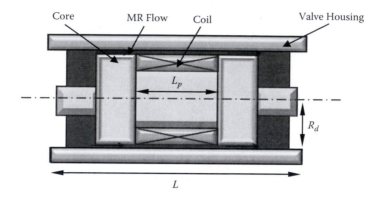

FIGURE 5.15
Schematic diagram of an MR damper.

Figure 5.15 shows the schematic configuration of a typical single-coil MR valve. The valve consists of valve coil, core, and housing. MR fluid flows from the inlet, through annular ducts between the core and the housing to the outlet. As the power of the coil is turned on, a magnetic field is exerted on the MR fluid in the annular duct between the magnetic poles. This causes the MR fluid to change its state into a semi-liquid or solid phase and hence stop the flow. Thus, this MR valve can be used as a relief valve, a pressure control valve, or a flow rate control valve.

By assuming Bingham fluid behavior of MR fluid and neglecting unsteady effect of MR fluid flow in the annular duct, the pressure drop of the single MR valve is calculated by

$$\Delta P_v = \Delta P_{vis} + \Delta P_y = \frac{6\eta L}{\pi t_d^3 R_d} Q + 2c \frac{L_p}{t_d} \tau_y \tag{5.38}$$

where ΔP_{vis} and ΔP_y are viscous and field-dependent pressure drop of the single annular MR valve, respectively. Q is the flow rate through the MR valve. τ_y and η are the induced yield stress and post-yield viscosity of the MR fluid. R_d is the average radius of the annular duct. L is the overall effective length. t_d is the annular duct gap and L_p is the magnetic pole length. The coefficient c depends on the MR flow velocity profile, which can be expressed by

$$c = 2.07 + \frac{12Q\eta}{12Q\eta + 0.8\pi R_d t_d^2 \tau_y} \tag{5.39}$$

The yield stress of the MR fluid, τ_y, is a function of the applied magnetic field intensity, which can be approximately expressed by

$$\tau_y = p(H_{mr}) = C_0 + C_1 H_{mr} + C_2 H_{mr}^2 + C_3 H_{mr}^3 \tag{5.40}$$

where H_{mr} is the applied magnetic field whose unit is A/mm. The magnetic field is related by $H_{mr} = NI/2t_d$. N is the coil turns of the MR valve and I is the current. The coefficient C is intrinsic value of the MR fluid to be experimentally determined. The pressure in the accumulator can be calculated as follows:

$$P_a = P_0 \left(\frac{V_0}{V_0 + A_s x_p} \right)^\gamma \tag{5.41}$$

where P_0 and V_0 are initial pressure and volume of the accumulator. γ is the coefficient of thermal expansion, which ranges from 1.4 to 1.7 for adiabatic expansion. δ_p is the piston displacement. It is noted that the piston displacement is relative displacement between car body and bogie frame of a railway vehicle. From Equation (5.36), Equation (5.37), and Equation (5.38), the damping force can be calculated by [12]

$$F_d = P_a A_s + c_{vis} \dot{x}_p + F_{MR} \, \text{sgn}(\dot{\delta}_p) \tag{5.42}$$

where

$$c_{vis} = \frac{12\eta L}{\pi R_d t_d^3} (A_p - A_s)^2$$

$$F_{MR} = (A_p - A_s) \frac{2cL_p}{t_d} \tau_y$$

The first term in Equation (5.42) represents the elastic force from the accumulator, the second term represents the damping force due to MR fluid viscosity, and the third term is the force due to the yield stress of the MR fluid that can be continuously controlled by the magnetic field across the MR fluid duct. In this work, a commercial product (MRF 132 DG) from Lord Corporation is used for the MR fluid. The coefficients C_0, C_1, C_2, C_3 are obtained from experimenting and determined using the least square curve fitting method: $C_0 = 0.3$, $C_1 = 0.42$, $C_2 = 1.16E3$, $C_3 = 1.05E - 6$. Based on the Bingham model of the MR fluid, the size and the level of required damping force are determined so that the MR damper can be applicable to a military vehicle. The principle design parameters of the MR damper are chosen as follows: the radius of MR valve (R_d): 28 mm; the length of the magnetic pole (L_p): 34 mm; the gap between the magnetic poles (t_d): 1 mm; and the number of coil turns (N): 280.

5.3.2 Vibration Control Results

In order to investigate the performance of the proposed MR damper, among many potential candidates for control algorithms, the semi-active skyhook controller is adopted in this work. It is well known that the logic of the skyhook is simple and easy to implement in a practical field. The desired damping force for the MR damper is set by

$$u = C_{sky} \cdot \dot{\delta} \tag{5.43}$$

where C_{sky} is the gain of the skyhook controller. The control gain physically indicates the damping coefficient. On the other hand, the damping force of the MR damper needs to be controlled depending upon the motion of the piston movement. Therefore, the following actuating condition is normally imposed.

$$u = \begin{bmatrix} C_{sky} \cdot \dot{\delta} & for\ \dot{\delta}(\dot{\delta}_{carbody} - \dot{\delta}_{bogieframe}) > 0 \\ 0, & for\ \dot{\delta}(\dot{\delta}_{carbody} - \dot{\delta}_{bogieframe}) \leq 0 \end{bmatrix} \tag{5.44}$$

The power spectral density (PSD) of lateral, yaw, and roll acceleration of the car body of a railway vehicle under random track irregularities is illustrated in Figure 5.16. The uncontrolled case represents the conventional passive system using viscous dampers without MR dampers for the secondary suspension system. For the controlled case, the MR dampers are operated in semi-active control mode via the proposed controller. From Figure 5.16, it is obvious that the secondary suspension system integrated with MR dampers is especially effective for reducing the lateral, yaw, and roll vibrations of the car body compared to the uncontrolled case. While the laterally installed MR dampers are effective for controlling the lateral, yaw, and roll vibrations of the car body, there is an approximate 24% to 30% improvement in terms of vibration reduction.

The dynamic performance of a railway vehicle as related to safety is evaluation in terms of specific performance indices, which include quantitative measurements of ride quality and vehicle stability. Ride quality is usually interpreted as the capability of the vehicle suspension to maintain the motion within the range of human comfort and within the range necessary to ensure that there is no loading damage. The ride quality of a vehicle depends on displacement, acceleration, rate of change of acceleration, and other factors such as noise, dust, humidity, and temperature. Generally, the approach used to evaluate the ride quality of a railway vehicle is the ride index method [13].

The ride index V_r is calculated from octave-band accelerations and the corresponding center frequency as follows:

$$V_r = \begin{cases} 8.1(a_{rms})^{0.3}(f_c/5.4)^{1/3}, & f_c < 5.4Hz \\ 8.1(a_{rms})^{0.3}(f_c/5.4), & f_c < 5.4Hz \end{cases} \tag{5.45}$$

where a_{rms} is the octave-bane acceleration and f_c is the octave-band center frequency. The ride indices for each frequency band are calculated and the overall ride index is obtained as follows:

$$V_r = \left(V_{r1}^{10} + V_{r2}^{10} + \cdots + V_{rn}^{10}\right)^{1/10} \tag{5.46}$$

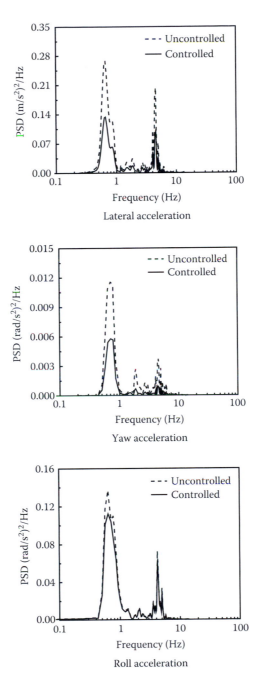

FIGURE 5.16
Power spectrum densities of the car body acceleration of the railway vehicle.

TABLE 5.4

Ride Quality Index

Ride Quality	Ride Index	
Very good	1	
Almost very good	1.5	
Good	2	
Almost good	2.5	
Satisfactory	3	
Just satisfactory	3.5	
Tolerable	4	
Intolerable	4.5	
Dangerous	5	
	Uncontrolled	**Controlled**
Lateral	2.6	1.4
Roll	2.8	2.3
Yaw	2.7	1.7

By using subjective rating, the ride indices were equated with ride quality as shown in Table 5.4. It is illustrated that the railway vehicle suspension system integrated with MR dampers is especially effective for increasing the lateral, yaw, and roll direction of the ride quality compared to the uncontrolled case.

5.4 Some Final Thoughts

In this chapter, we have seen that the MR suspension systems can be successfully adopted for not only passenger vehicles but also tracked vehicles and railway vehicles in order to attenuate vibrations from road conditions.

In Section 5.2, an MRSU was proposed for tracked vehicles. A double-rod-type MRSU was devised and its damping characteristics and spring characteristics were analyzed. An optimization procedure based on the finite element analysis has been developed in order to find the optimal geometry of an MR valve, which is the key component of the MRSU. Optimal solutions have been obtained and the performance of the MRSU such as damping torque has been analyzed and presented. Then a tracked vehicle installed with the MRSU was modeled considering vertical motion. It has been shown through computer simulation that in a bump test the vertical motion can be significantly reduced by employing the proposed MRSU associated with the skyhook controller. In addition, it has been demonstrated that the optimized MRSU can provide better vibration control performance than the initial one with smaller control energy.

In Section 5.3, a semi-active secondary suspension system with MR dampers was investigated for railway vehicles. A 15 degree-of-freedom railway vehicle model was derived, which includes three vibration motions such as lateral, yaw, and roll of the wheel-set, bogie frame, and car body. The governing equations of the vehicle motion were then integrated with MR dampers. Subsequently, in order to demonstrate the effectiveness of the MR damper, computer simulations were undertaken via skyhook control algorithm under random irregularity using creep force. It is assured from the results that ride quality of the railway vehicle can be improved substantially by employing the proposed MR damper.

References

[1] Wang, W., Chen, B., Yan, X., and Gu, L. 2011. Design and dynamic simulation on adjustable stiffness hydro-pneumatic suspension vibration characters of tracked vehicle. *Electric Information and Control Engineering (ICEICE)*, Beijing, China, pp. 4056–4059.

[2] Kamath, G. M. and Wereley, N. M. 1997. Nonlinear viscoelastic plastic mechanism based model of an electrorheological damper. *Smart Materials and Structures*. 6: 351–359.

[3] Gavin, H. P., Hanson, R. D., and Filisko, F. E. 1996. Electrorheological dampers, part 2: testing modelling. *Journal of Applied Mechanics*. 63: 676–682.

[4] Choi, S. B., Choi, Y. T., Chang, E. G., Han, S. J., and Kim, C. S. 1998. Control characteristic of a continuously variable ER damper. *Mechatronics*. 8: 570–576.

[5] Ha, S. H., Choi, S. B., Rhee, E. J., and Kang, P. S. 2009. Optimal design of a magnetorheological fluid suspension for tracked vehicle. *Journal of Physics: Conference Series*.149: 1–5.

[6] Iwnicki, S. 2006. *Handbook of Railway Vehicle Dynamics*. Boca Raton: CRC/Taylor & Francis.

[7] Ha, S. H., Choi, S. B., Lee, K. S., and Cho, M. W. 2011. Ride quality evaluation of railway vehicle suspension system featured by magnetorhological fluid damper. *2011 International Conference on Mechatronics and Materials Processing (ICMMP2011)*, Guangzhou, China.

[8] Ha, S. H., Seong, M. S., Kim, H. S., and Choi, S. B. 2011. Performance evaluation of railway secondary suspension utilizing magnetorhological fluid damper. *The 12th International Conference on Electrorheological (ER) Fluids and Magnetorheological (MR) Suspensions*, Philadelphia, PA, pp. 142–148.

[9] Coodall, R. M., Williams, R. A., Lawton, A. and Harborough, P. R. Railway Vehicle Active Suspensions in Theory and Practice. 1981. *Vehicle System Dynamics: International Journal of Vehicle Mechanics and Mobility* 10: 210–215.

[10] Choi, S. B., Shu, M. S., Park, D. W., and Shin M. J. 2001. Neuro-fuzzy control of a tracked vehicle featuring semi-active electro-rheological suspension units. *Vehicle System Dynamics* 35: 141–162.

[11] Liao, W. H. and Wang, D. H. Semiactive vibration control of train suspension systems via magnetorheological dampers. 2003. *Journal of Intelligent Material Systems and Structures* 14: 161–172.

[12] Nguyen, Q. H., Han, Y. M., Choi, S. B., and Wereley, N. M. 2007. Geometry optimization of MR valves constrained in a specific volume using finite element method. *Smart Materials and Structures*. 16: 2242–2252.

[13] Coxon, H. E. and McHaughton, L. D. 1971. The elements of bogie design for Australian conditions, NSW Railways–Institution of Mechanical Engineers (Australia).

6

MR Applications for Vibration and Impact Control

6.1 Introduction

Chapters 4 and 5 discussed the rheological properties of MR fluids, which are controllable by magnetic field intensity and are very effective for vibration control systems of vehicle bodies such as a semi-active suspension system adopting MR shock absorbers. Vehicles have several vibration or impact sources, such as engine excitation or external collision. In this case, the vehicle body vibration was caused by road excitation.

Engine excitation is one serious vibration source for passenger vehicles. Therefore, various types of passive engine mounts have been proposed and developed to attenuate unwanted engine noise and vibration [1, 2]. Controllable engine mounts have recently been proposed, and their superior vibration control performances compared to conventional passive engine mounts have been demonstrated via analytic and empirical approaches [3–7]. One attractive approach to attenuating unwanted vibration due to engine excitation is to utilize a semi-active engine mount that adopts MR fluids.

Vehicle collisions can injure a driver and damage a vehicle body. It is essential that the driver and vehicle body be secured against vehicle collisions. Accordingly, research is actively focusing on brake systems that induce braking force in advance. This improves the safety of the system that protects vehicles and drivers from collisions. Much research has been performed on the safety of drivers who use air bags and safety belts as well as the reduced collision rates resulting from braking efficiency improvements such as anti-lock brake systems [8]. Mizuno and Kajzer tried to observe both collision characteristics from various directions by car type and collision accident analyses of pedestrians [9]. Witteman proposed a friction damper to guarantee vehicle security by reducing impact force and demonstrated it via a vehicle model analysis using the finite element method (FEM) [10]. Wagstrom, Thomson, and Pipkorn observed that a vehicle stiffness change could improve stability in various conditions through a vehicle model that combined a 1 degree-of-freedom system using a computer simulation [11].

Jawad considered a smart bumper before collision controller that could change stiffness prior to a collision utilizing a hydraulic system [12]. Research on shock reduction was recently accomplished using smart fluid, which has reversible properties under an applied field. Lee, Choi, and Wereley suggested the use of an MR damper to reduce the shock transmitted to a helicopter. They then designed the controller using mathematic modeling and verified its performance by applying it to an actual system [13]. Ahmadian, Appleton, and Norris suggested a shock reduction mechanism that used an MR damper when a bomb is fired [14]. Lee and Choi [15] and Lee, Choi, and Lee [16] verified controller performance using an MR damper applied to a vehicle suspension. Song et al. proposed use of a shock damper to reduce impact and then substantiated the impact control performance using decrements of acceleration as a 1 degree-of-freedom system [17].

Section 6.2 introduces a mixed-mode (combination of shear and flow modes) MR engine mount to effectively attenuate unwanted vibration of a passenger vehicle [18]. To achieve this goal, an engine mount is first designed and manufactured by adopting a mixed-mode configuration and MR fluids. After verification that the damping force is controlled by magnetic field (or current) intensity tuning, the MR engine mount is incorporated into a full-vehicle model. The governing equations of motion are derived under consideration of engine excitation force. A semi-active skyhook controller is then formulated and implemented using a hardware-in-the-loop simulation (HILS). Vibration control responses at the driver's position are evaluated in both frequency and time domains.

Section 6.3 introduces an MR impact damper that reduces the amount of force transmitted into a vehicle chassis during a frontal vehicle collision [19]. The governing equations of motion of the MR impact damper are derived using a hydraulic model. The dynamic equation includes a field-dependent force that is achieved from a Bingham model of the MR fluid. The collision mitigation performance is then evaluated using a computer simulation that involves a vehicle model with an occupant.

6.2 MR Engine Mount

6.2.1 Configuration and Modeling

The schematic configuration of the mixed-mode MR engine mount is shown in Figure 6.1(a). The upper part of the mount consists of main rubber in order to provide appropriate stiffness and damping properties. The magnetic pole is fixed to the upper plate, and therefore it is moved with the rubber in the vertical direction depending upon the external excitation. The MR fluid motion is induced through a gap between the housing and the magnetic pole, and controlled by the intensity of the magnetic field (or current) applied to

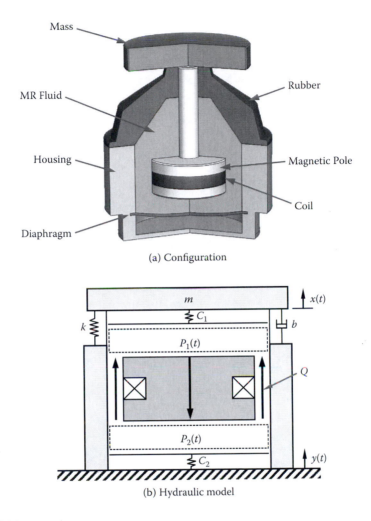

FIGURE 6.1
The proposed mixed-mode MR engine mount.

the coil. Thus, the damping force of the mount can be controlled. By assuming that the MR fluid is incompressible and the pressure in the chamber is uniform, the dynamic equation of the MR engine mount is derived from the hydraulic model shown in Figure 6.1(b) and given by

$$m\ddot{x}(t) = -k(x(t) - y(t)) - b(\dot{x}(t) - \dot{y}(t)) - A\eta\frac{\dot{x}(t) - \dot{y}(t)}{h}$$

$$+ A_p P_1(t) - A\tau_{ys}(H)\mathrm{sgn}(\dot{x}(t) - \dot{y}(t)) \tag{6.1}$$

where m is the engine mass, k is the stiffness of the rubber, b is the damping constant of the rubber, A is the flow area, h is the gap of the magnetic pole,

A_p is the piston area of the upper chamber, η is the viscosity of the MR fluid, and τ_{ys} is the yield stress of the MR fluid under the shear mode. The pressure drop between the upper and the lower chambers is obtained by [20]:

$$P_2(t) - P_1(t) = \Delta P_\eta(t) + \Delta P_{MR}(t)$$

$$= \left(\frac{2n+1}{n}\frac{2Q_f(t)}{wh^2}\right)^n \frac{2\eta L}{h} + D\frac{2L\tau_{yf}(H)\operatorname{sgn}(Q_f(t))}{h} \tag{6.2}$$

where

$$D = \frac{2n+1}{n+1} - \frac{3n(1+2n)(1-n)}{16(n+1)^2} - \frac{3n(n^2+1)}{40(n+1)}$$

In the above, Q_f is the fluid flow resulting from pressure difference, τ_{yf} is the yield stress of the MR fluid under the flow mode, n is the flow behavior index of the Herschel-Bulkley model [21], L is the length of the magnetic pole, and w is the width of the magnetic pole. The pressure P_1 and P_2 in Equation (6.2) can be obtained from the following continuity equations [22]:

$$C_1\dot{P}_1(t) = Q(t) - A_p(\dot{x}(t) - \dot{y}(t))$$

$$C_2\dot{P}_2(t) = -Q(t) \tag{6.3}$$

$$Q(t) = Q_f(t) + A_s(\dot{x}(t) - \dot{y}(t))/2$$

where Q is the total fluid flow, and C_1 and C_2 are the compliance of the upper and lower chambers, respectively.

From Equation (6.2) and Equation (6.3), we can obtain the following equation:

$$Q_f^{n-1}(t)\dot{Q}_f(t) = \left(-\frac{C_1+C_2}{C_1C_2R}Q(t) + \frac{A_p}{C_1R}(\dot{x}(t) - \dot{y}(t)) - \frac{1}{R}\frac{d}{dt}\frac{D2L\tau_{yf}(H)\operatorname{sgn}(Q_f(t))}{h}\right)$$

$$\tag{6.4}$$

where R is the flow resistance in the absence of the magnetic field. Since $(C_1+C_2)/(C_1C_2R) \gg 1$, we can assume $\dot{Q}_f = 0$ [23]. In addition, since the response of the pressure drop due to the magnetic field is much faster than the exciting frequency, the derivative term (third term in Equation (6.4)) can be neglected. Thus, the governing equation of the MR engine mount is obtained as:

$$m\ddot{x}(t) + \left\{b - A\frac{\eta}{h}\right\}(\dot{x}(t) - \dot{y}(t)) + \left\{k + \frac{A_p^2}{C_1+C_2}\right\}(x(t) - y(t))$$

$$+ \frac{C_2 A_P}{C_1+C_2}\left\{\frac{2n+1}{n}\frac{2}{wh^2}\left(\frac{C_2 A_P}{C_1+C_2} - \frac{A}{2}\right)(\dot{x}(t) - \dot{y}(t))\right\}^n \frac{2\eta L}{h} = -F_{MR} \tag{6.5}$$

FIGURE 6.2
The photograph of the MR engine mount.

where

$$F_{MR} = \left(\frac{A_P C_2}{C_1 + C_2} D \frac{2L\tau_{yf}(H)}{h} + A\tau_{ys}(H) \right) \text{sgn}(\dot{x}(t) - \dot{y}(t))$$

Based on the governing model given by Equation (6.5), an appropriate size of the MR engine mount is designed and manufactured as shown in Figure 6.2. The damping force level of the MR engine mount designed in this test is applicable to a middle-sized passenger vehicle. In order to investigate the field-dependent dynamic performance of the MR engine mount, an experimental apparatus shown in Figure 6.3 is established. An engine mass of 60 kg is imposed and the displacement transmissibility between input excitation and output displacement is evaluated at various magnetic fields. The displacements of the input and output are measured by two sets of proximitors and various magnetic fields are induced by applying various current inputs to the current amplifier. Figure 6.4 presents the measured transmissibility at two different excitation levels. It is clearly observed that the proposed MR engine mount effectively suppresses the vibration at the neighborhood of resonance. This directly indicates that the damping force of the MR engine mount can be controlled by controlling the intensity of the current. In order to verify the damping force controllability, a semi-active skyhook controller is designed and experimentally implemented. Figure 6.5 presents the measured damping force controllability of the MR engine mount. We clearly see that the damping force is produced only by the viscosity of the MR

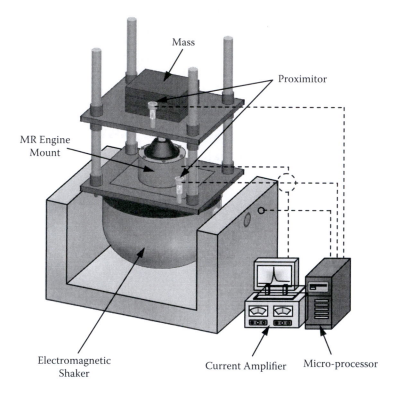

FIGURE 6.3
An experimental setup for the MR engine mount test.

fluid before control action. However, the desired damping force imposed in the microprocessor is well tracked by activating the MR engine mount associated with the skyhook controller. This excellent advantage of the damping force controllability will provide an effective vibration control of a full-vehicle system.

6.2.2 Full-Vehicle Model

Let us consider the 13 degrees-of-freedom full-vehicle model shown in Figure 6.6. The vehicle is front-engine-front-drive (FF) type and the engine is supported by three engine mounts. The first and second MR engine mounts are situated at the front and back left corner, respectively, and the third MR engine mount is installed at the right edge of the engine in the (x,z) plane. The vehicle body is assumed rigid, while the engine is assumed to be having degrees of freedom in pitch, roll, and yaw directions. In addition, it is assumed that the vehicle body is connected to four rigid bodies representing wheel unsprung mass, each having vertical degrees of

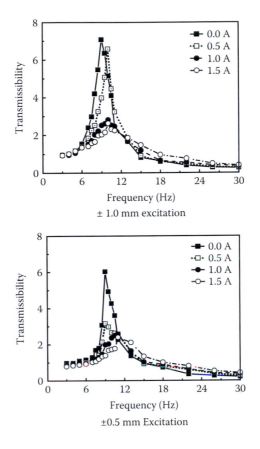

FIGURE 6.4
Displacement transmissibility of the MR engine mount. (From Choi, S.B. et al., *International Journal of Vehicle Design*, 33, 1-3, 2003. With permission.)

freedom. The governing equations of motion of the full-vehicle model are derived as [20]:

1. Vertical motion(z_b) of the vehicle body:

$$m_b \ddot{z}_b = -k_{s1}(z_{b1} - z_{u1}) - c_{s1}(\dot{z}_{b1} - \dot{z}_{u1}) - k_{s2}(z_{b2} - z_{u2}) - c_{s2}(\dot{z}_{b2} - \dot{z}_{u2})$$

$$- k_{s3}(z_{b3} - z_{u3}) - c_{s3}(\dot{z}_{b3} - \dot{z}_{u3}) - k_{s4}(z_{b4} - z_{u4}) - c_{s4}(\dot{z}_{b4} - \dot{z}_{u4})$$

$$+ k_{e1Z}(Z_{e1} - Z_{eb1}) + c_{e1Z}(\dot{Z}_{e1} - \dot{Z}_{eb1}) + k_{e2Z}(Z_{e2} - Z_{eb2}) + c_{e2Z}(\dot{Z}_{e2} - \dot{Z}_{eb2})$$

$$+ k_{e3Z}(Z_{e3} - Z_{eb3}) + c_{e3Z}(\dot{Z}_{e3} - \dot{Z}_{eb3}) + F_{MR1Z} + F_{MR2Z} + F_{MR3Z}$$

$$(6.6)$$

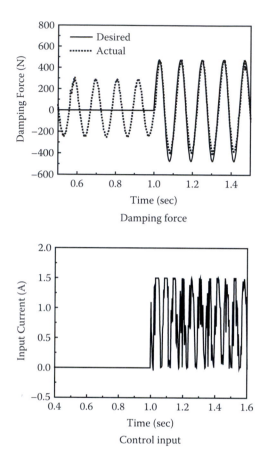

FIGURE 6.5
Damping force controllability of the MR engine mount. (From Choi, S.B. et al., *International Journal of Vehicle Design*, 33, 1-3, 2003. With permission.)

2. Roll motion(ϕ) of the vehicle body:

$$J_\phi \ddot{\phi} = -t_{br}(k_{s1}(z_{b1} - z_{u1}) + c_{s1}(\dot{z}_{b1} - \dot{z}_{u1})) - t_{br}(k_{s2}(z_{b2} - z_{u2}) + c_{s2}(\dot{z}_{b2} - \dot{z}_{u2}))$$
$$+ t_{bl}(k_{s3}(z_{b3} - z_{u3}) + c_{s3}(\dot{z}_{b3} - \dot{z}_{u3})) + t_{bl}(k_{s4}(z_{b4} - z_{u4}) + c_{s4}(\dot{z}_{b4} - \dot{z}_{u4}))$$
$$- t_{bel}(k_{e1Z}(Z_{e1} - Z_{eb1}) + c_{e1Z}(\dot{Z}_{e1} - \dot{Z}_{eb1})) - t_{bel}(k_{e2Z}(Z_{e2} - Z_{eb2})$$
$$+ c_{e2Z}(\dot{Z}_{e2} - \dot{Z}_{eb2})) + t_{ber}(k_{e3Z}(Z_{e3} - Z_{eb3}) + c_{e3Z}(\dot{Z}_{e3} - \dot{Z}_{eb3}))$$
$$- t_{bel}F_{MR1Z} - t_{bel}F_{MR2Z} + t_{ber}F_{MR3Z}$$

(6.7)

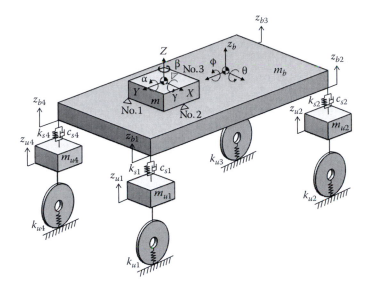

FIGURE 6.6
Full-vehicle model with the MR engine mounts.

3. Pitch motion(θ) of the vehicle body:

$$J_\theta \ddot{\theta} = l_{bf}(k_{s1}(z_{b1} - z_{u1}) + c_{s1}(\dot{z}_{b1} - \dot{z}_{u1})) - l_{br}(k_{s2}(z_{b2} - z_{u2}) + c_{s2}(\dot{z}_{b2} - \dot{z}_{u2}))$$
$$- l_{br}(k_{s3}(z_{b3} - z_{u3}) + c_{s3}(\dot{z}_{b3} - \dot{z}_{u3})) + l_{bf}(k_{s4}(z_{b4} - z_{u4}) + c_{s4}(\dot{z}_{b4} - \dot{z}_{u4}))$$
$$- l_{bef}(k_{e1Z}(Z_{e1} - Z_{eb1}) + c_{e1Z}(\dot{Z}_{e1} - \dot{Z}_{eb1})) - l_{ber}(k_{e2Z}(Z_{e2} - Z_{eb2}) \tag{6.8}$$
$$+ c_{e2Z}(\dot{Z}_{e2} - \dot{Z}_{eb2})) - l_{bem}(k_{e3Z}(Z_{e3} - Z_{eb3}) + c_{e3Z}(\dot{Z}_{e3} - \dot{Z}_{eb3}))$$
$$- l_{bef} F_{MR1Z} - l_{ber} F_{MR2Z} - l_{bem} F_{MR3Z}$$

4. Suspension motion:

$$m_{u1}\ddot{z}_{u1} = k_{s1}(z_{b1} - z_{u1}) + c_{s1}(\dot{z}_{b1} - \dot{z}_{u1}) - k_{u1}z_{u1}$$
$$m_{u2}\ddot{z}_{u2} = k_{s2}(z_{b2} - z_{u2}) + c_{s2}(\dot{z}_{b2} - \dot{z}_{u2}) - k_{u2}z_{u2}$$
$$m_{u3}\ddot{z}_{u3} = k_{s3}(z_{b3} - z_{u3}) + c_{s3}(\dot{z}_{b3} - \dot{z}_{u3}) - k_{u3}z_{u3} \tag{6.9}$$
$$m_{u4}\ddot{z}_{u4} = k_{s4}(z_{b4} - z_{u4}) + c_{s4}(\dot{z}_{b4} - \dot{z}_{u4}) - k_{u4}z_{u4}$$

5. Engine motion in Z direction:

$$m_e\ddot{Z} = -k_{e1Z}(Z_{e1} - Z_{eb1}) - c_{e1Z}(\dot{Z}_{e1} - \dot{Z}_{eb1}) - k_{e2Z}(Z_{e2} - Z_{eb2})$$
$$- c_{e2Z}(\dot{Z}_{e2} - \dot{Z}_{eb2}) - k_{e3Z}(Z_{e3} - Z_{eb3}) - c_{e3Z}(\dot{Z}_{e3} - \dot{Z}_{eb3}) \tag{6.10}$$
$$- F_{MR1Z} - F_{MR2Z} - F_{MR3Z} + F_{eZ}$$

6. Engine motion in X direction:

$$m_e\ddot{X} = -k_{e1X}(X_{e1} - X_{eb1}) - c_{e1X}(\dot{X}_{e1} - \dot{X}_{eb1}) - k_{e2X}(X_{e2} - X_{eb2})$$
$$- c_{e2X}(\dot{X}_{e2} - \dot{X}_{eb2}) - k_{e3X}(X_{e3} - X_{eb3}) - c_{e3X}(\dot{X}_{e3} - \dot{X}_{eb3}) + F_{eX} \quad (6.11)$$

7. Engine motion in Y direction:

$$m_e\ddot{Y} = -k_{e1Y}(Y_{e1} - Y_{eb1}) - c_{e1Y}(\dot{Y}_{e1} - \dot{Y}_{eb1}) - k_{e2Y}(Y_{e2} - Y_{eb2})$$
$$- c_{e2Y}(\dot{Y}_{e2} - \dot{Y}_{eb2}) - k_{e3Y}(Y_{e3} - Y_{eb3}) - c_{e3Y}(\dot{Y}_{e3} - \dot{Y}_{eb3}) + F_{eY} \quad (6.12)$$

8. Rolling motion (α) of the engine:

$$J_\alpha\ddot{\alpha} = +l_{ef}(k_{e1Z}(Z_{e1} - Z_{eb1}) + c_{e1Z}(\dot{Z}_{e1} - \dot{Z}_{eb1}))$$
$$+ h_e(k_{e1Y}(Y_{e1} - Y_{eb1}) + c_{e1Y}(\dot{Y}_{e1} - \dot{Y}_{eb1}))$$
$$- l_{er}(k_{e2Z}(Z_{e2} - Z_{eb2}) + c_{e2Z}(\dot{Z}_{e2} - \dot{Z}_{eb2}))$$
$$+ h_e(k_{e2Y}(Y_{e2} - Y_{eb2}) + c_{e2Y}(\dot{Y}_{e2} - \dot{Y}_{eb2}))$$
$$+ l_{ef}F_{MR1Z} - l_{er}F_{MR2Z} + M_{eX} \quad (6.13)$$

9. Yawing motion (β) of the engine:

$$J_\beta\ddot{\beta} = -l_{ef}(k_{e1X}(X_{e1} - X_{eb1}) + c_{e1X}(\dot{X}_{e1} - \dot{X}_{eb1}))$$
$$- t_{el}(k_{e1Y}(Y_{e1} - Y_{eb1}) + c_{e1Y}(\dot{Y}_{e1} - \dot{Y}_{eb1}))$$
$$+ l_{er}(k_{e2X}(X_{e2} - X_{eb2}) + c_{e2X}(\dot{X}_{e2} - \dot{X}_{eb2}))$$
$$- t_{el}(k_{e2Y}(Y_{e2} - Y_{eb2}) + c_{e2Y}(\dot{Y}_{e2} - \dot{Y}_{eb2}))$$
$$+ t_{er}(k_{e3Y}(Y_{e3} - Y_{eb3}) + c_{e3Y}(\dot{Y}_{e3} - \dot{Y}_{eb3})) + M_{eZ} \quad (6.14)$$

10. Pitching motion(γ) of the engine:

$$J_\gamma\ddot{\gamma} = -h_e(k_{e1X}(X_{e1} - X_{eb1}) + c_{e1X}(\dot{X}_{e1} - \dot{X}_{eb1}))$$
$$+ t_{el}(k_{e1Z}(Z_{e1} - Z_{eb1}) + c_{e1Z}(\dot{Z}_{e1} - \dot{Z}_{eb1}))$$
$$- h_e(k_{e2X}(X_{e2} - X_{eb2}) + c_{e2X}(\dot{X}_{e2} - \dot{X}_{eb2}))$$
$$+ t_{el}(k_{e2Z}(Z_{e2} - Z_{eb2}) + c_{e2Z}(\dot{Z}_{e2} - \dot{Z}_{eb2}))$$
$$- t_{er}(k_{e3Z}(Z_{e3} - Z_{eb3}) + c_{e3Z}(\dot{Z}_{e3} - \dot{Z}_{eb3}))$$
$$+ t_{el}F_{MR1Z} + t_{el}F_{MR2Z} - t_{er}F_{MR3Z} + M_{eY} \quad (6.15)$$

TABLE 6.1

Specifications of the Vehicle Parameters

Variable	Value	Unit	Variable	Value	Unit
m_b	868	kg	k_{e1}, k_{e2}, k_{e3}	133–240	N/m
m_e	244	kg	t_{el}	0.25	m
m_{u1}, m_{u4}	29.5	kg	t_{er}	0.52	m
m_{u2}, m_{u3}	27.5	kg	l_{ef}	0.19	m
c_{s1}, c_{s4}	3200	N200e	l_{er}	0.21	m
c_{s2}, c_{s3}	1700	N700e	h_e	0.16	m
$k_{u1}, k_{u2}, k_{u3}, k_{u4}$	200,000	N/m	l_{bef}	1.3	m
k_{s1}, k_{s4}	20,580	N/m	l_{bem}	1.11	m
k_{s2}, k_{s3}	19,600	N/m	l_{ber}	0.9	m
J_ϕ	235	kg51²	t_{bl}	0.72	m
J_θ	920	kg02²	t_{br}	0.72	m
J_α	25	kg72²	l_{bf}	1.4	m
J_γ	34	kg42²	l_{br}	1.4	m
J_β	30	kg42²	l_c	0.01	m
c_{e1}, c_{e2}, c_{e3}	610	N101e	l_r	0.135	m

The variables used from Equation (6.6) to Equation (6.15) are well defined with specific values in Table 6.1.

Now, in order to solve the equations of the motion, we have to determine engine excitation force. These are given by [24, 25]:

$$F_{eX} = 0$$

$$F_{eY} = 0$$

$$F_{eZ} = \frac{4m_p r^2 \omega^2}{l} \cos 2\omega t$$

$$M_{eX} = \sum_{n=1}^{4} P_{cn} A_c \left(r \sin(\omega t - \phi_{gn}) + \frac{r^2}{2l} \sin 2(\omega t - \phi_{gn}) \right)$$

$$+ \left(-2m_p r^2 \omega^2 \sin 2\omega t + F_{eZ} l_c \right)$$

$$M_{eY} = \frac{4m_p r^2 \omega^2}{l} l_r \cos 2\omega t$$

$$M_{eZ} = 0$$

(6.16)

In the above, m_p is the piston mass, r is the rotational radius of crank arm, ω is the angular velocity of crank shaft, l is the connecting rod length, P_{cn} is the gas pressure in each cylinder, A_c is the piston area of the engine cylinder, and ϕ_{gn} is the gas explosion phase between each cylinder. The engine type adopted in this test is in-line, 4 cylinders, and 4 strokes.

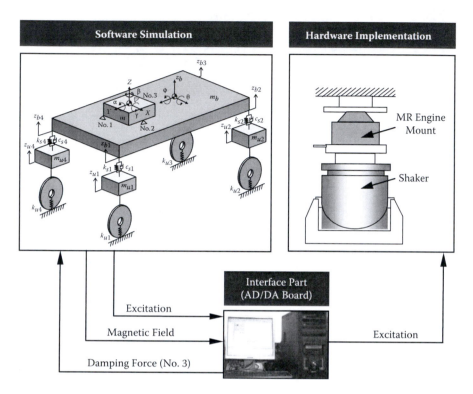

FIGURE 6.7
Schematic configuration of the HILS for vibration control.

6.2.3 Control Responses

Figure 6.7 shows an HILS configuration to evaluate vibration control performance of the MR engine mount. The HILS system consists of a software part, a hardware part, and an interface part. The software part is the full-vehicle model, while the hardware part is featured by shaker, load cell, proximitor, and MR engine mount. The interface part is composed of a microprocessor with a digital signal processing (DSP) board, and A/D and D/A converters. The damping force of the MR engine mount is experimentally measured and applied to the corresponding component of the full-vehicle model. The vibration control response of the vehicle is then evaluated and the shaker excites the engine mount again depending upon the controlled motion. This looping operation is continuously undertaken until a favorable control response is achieved.

In this test, the magnetic field (or current) to be applied to the MR engine mount is determined by adopting the semi-active skyhook controller [26]. The desired damping force at each position of the MR engine mount is set by

$$U_i(t) = C_{skyi}\left(\dot{Z}_{ei}(t) - \dot{Z}_{ebi}(t)\right) = F_{MRi}(t) \qquad (i = 1, 2, 3) \qquad (6.17)$$

where C_{skyi} is control gain, which physically indicates the damping coefficient. The damping force set by Equation (6.17) needs to be employed to the MR engine mount depending upon the relative motion. Thus, the following actuating condition is imposed:

$$U_i(t) = \begin{bmatrix} U_i(t) & for\ \dot{Z}_{ei}(t)(\dot{Z}_{ei}(t) - \dot{Z}_{ebi}(t)) > 0 \\ 0 & for\ \dot{Z}_{ei}(t)(\dot{Z}_{ei}(t) - \dot{Z}_{ebi}(t)) \leq 0 \end{bmatrix} \tag{6.18}$$

The above condition physically implies that the control U_i only assures the increment of energy dissipation of the stable system [27]. Once the control input U_i is determined, the input magnetic field to be applied to the MR engine mount is obtained by

$$H_i(t) = \frac{-\alpha_2 + \sqrt{\alpha_2^2 + 4\alpha_1 \dfrac{C_1 + C_2}{A_pC_2} \dfrac{hU_i(t)}{2LD}}}{2\alpha_1} \tag{6.19}$$

where α_1 and α_2 are experimental constants of the field-dependent yield stress of the MR fluid given by $\tau_y(H) = \alpha_1 H^2 + \alpha_2 H$. In this test, the product of MRF-132LD (Lord Corp.) is used for the MR fluid, and α_1 and α_2 are experimentally evaluated as 3.16 and 224.25, respectively. It is noted that the control magnetic field obtained by Equation (6.19) can be converted easily to the control input current by considering the number of coil turns. It is also remarked that Equation (6.19) is obtained by assuming that the yield stress of the MR fluid in the flow mode is much greater than that in the shear mode [20].

Prior to evaluating vibration control performance, the forces transmitted from the engine to the vehicle body are evaluated and presented in Figure 6.8. It is clearly observed that the transmitted forces are favorably reduced by controlling the damping force of the MR engine mount. These results directly lead to the suppression of the vibration of the engine itself and the vehicle body. It is remarked that uncontrolled responses in the results are obtained in the absence of the control current. Figure 6.9 and Figure 6.10 present the frequency responses of the displacement and acceleration at the engine C.G. (center of gravity) point, respectively. We clearly see that vibration levels are well suppressed in the considered frequency range by actuating the MR engine mount. In order to explicitly observe the level of the vertical acceleration, the time history of the acceleration at the driver's position is evaluated and presented in Figure 6.11. The response is obtained by exciting the engine with an idle speed of 750 rpm. It is obvious that the magnitude of the acceleration is favorably suppressed by employing the control input current shown in Figure 6.11(b).

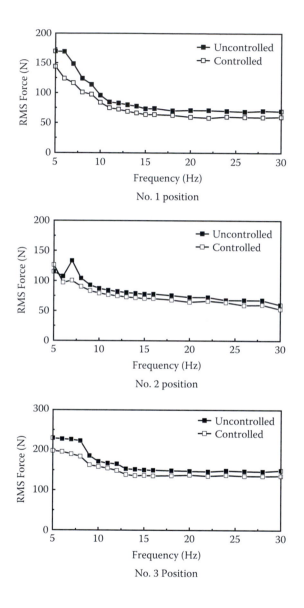

FIGURE 6.8
Force transmitted from the engine to the vehicle body. (From Choi, S.B. et al., *International Journal of Vehicle Design*, 33, 1-3, 2003. With permission.)

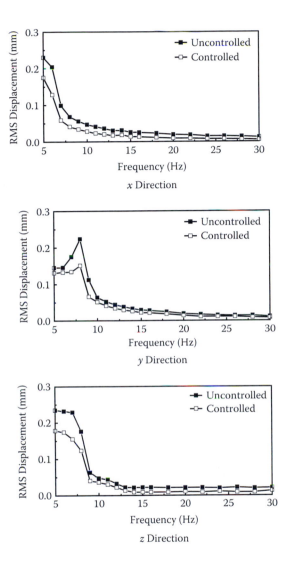

FIGURE 6.9
Root mean square (RMS) displacement at engine C.G. point. (From Choi, S.B. et al., *International Journal of Vehicle Design*, 33, 1–3, 2003. With permission.)

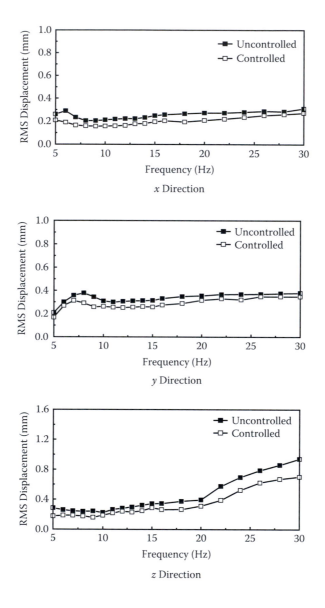

FIGURE 6.10
RMS acceleration at engine C.G. point. (From Choi, S.B. et al., *International Journal of Vehicle Design*, 33, 1-3, 2003. With permission.)

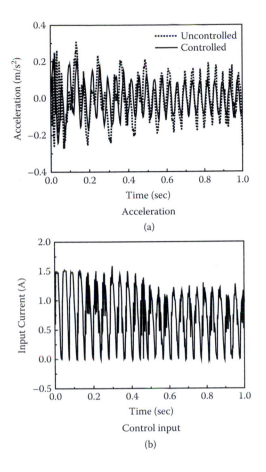

FIGURE 6.11

Acceleration at driver's position (750 rpm). (From Choi, S.B. et al., *International Journal of Vehicle Design*, 33, 1-3, 2003. With permission.)

6.3 MR Impact Damper

6.3.1 Dynamic Modeling

Figure 6.12 shows the configuration of an impact damper, which can be installed in a bumper structure of a vehicle. In this section, the MR fluid is introduced to generate damping force when a certain external impact occurs such as in a frontal collision. The MR impact damper consists of an upper chamber filled with the MR fluid, a magnetic circuit to produce a magnetic field, a lower chamber located under the magnetic circuit, and a diaphragm. The diaphragm contains and accumulates the fluid. A magnetic circuit is installed at the inner cylinder. A flow channel is located between the upper chamber and the lower chamber. After impact, the

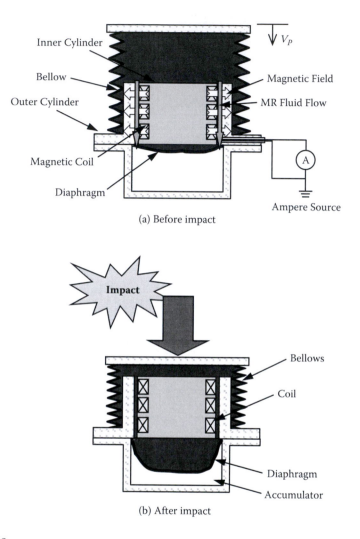

FIGURE 6.12
Configuration of the MR impact damper.

damper deforms as shown in Figure 6.12(b). The flow through the flow channel is generated by the deformation of the upper chamber, which has a bellows structure. The bellows have good properties of large deformation and fluid volume change. It has oil resistance, moisture resistance, pressure resistance, and airtight characteristics. When a magnetic field is applied during impact, the MR impact damper produces damping force at the magnetic pole. It can be generated continuously by adjusting the intensity of the magnetic field. In advance, a conventional finite element analysis program, ANSYS, is used to calculate magnetic flux density inside the flow channel. Figure 6.13 shows that the flow channel has a uniform magnetic

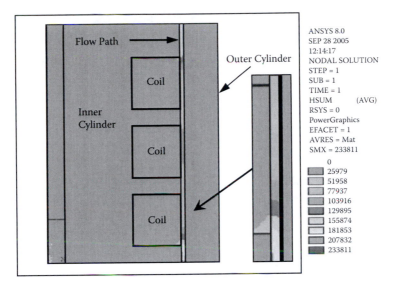

FIGURE 6.13
FEM result of the magnetic flux. (From Woo, D. et al., *Journal of Intelligent Material Systems and Structures*, 18, 12, 2007. With permission.)

flux. The dimensions of the flow channel and the magnetic pole are then determined based on the calculated magnetic flux density. It is assumed for analysis that the MR fluid is incompressible, and the MR impact damper has a motion in one direction.

Figure 6.14 shows a hydraulic model of the MR impact damper. The transmitted force into the chassis body is obtained as:

$$F_{cb} = -(k_b x(t) + c_b \dot{x}(t) - A_p P_1(t)) \qquad (6.20)$$

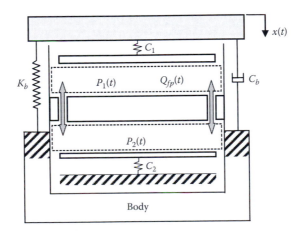

FIGURE 6.14
Hydraulic model of the MR Impact damper.

where k_b and c_b are the stiffness and damping constant of the bellows, P_1 is the pressure of the upper chamber, A_p is the effective piston area of the upper chamber, and $A_p P_1(t)$ is the produced damping force caused by pressure drop. The pressure drop consists of fluid resistance ΔP_{fp} and yield stress ΔP_{MR} of the MR fluid. The fluid resistance is generated by fluid flow without a magnetic field. The yield stress is produced by field intensity between magnetic poles. The pressure drop is derived as:

$$P_2(t) - P_1(t) = \Delta P_{fp} + \Delta P_{MR} \tag{6.21}$$

where ΔP_{fp} is represented as $R_{fp} Q_{fp}(t)$. R_{fp} is the fluid resistance and $Q_{fp}(t)$ is the flow rate defined as $A_{fp}\dot{x}_{fp}(t)$. A_{fp} is the area of the flow path and $\dot{x}_{fp}(t)$ is the flow velocity. ΔP_{MR} is represented as $2L/h \cdot \tau_y(H)$. L is the length of the magnetic pole (50 mm). h and H are the gap size (1 mm) and magnetic field, respectively.

When the flow is generated by external excitation $x(t)$, continuity equations at the upper and lower chambers are obtained as:

$$C_1 \dot{P}_1(t) = Q_{fp}(t) - A_p \dot{x}(t)$$
$$C_2 \dot{P}_2(t) = -Q_{fp}(t) \tag{6.22}$$

where C_1 and C_2 are the compliances of the upper and lower chambers. From Equation (6.20) to Equation (6.22), the transmitted force to body is represented as:

$$F_{cb} = -k_b x(t) - c_b \dot{x}(t) - k_i(x(t) - x_{fp}(t))$$
$$k_i(x(t) - x_{fp}(t)) - k_d x_{fp}(t) = c_{fp}\dot{x}_{fp}(t) + \text{sgn}(\dot{x}_{fp}(t))F_{MR} \tag{6.23}$$

where

$$k_i \equiv A_p^2/C_1, \, k_d \equiv A_p^2/C_2, \, c_{fp} \equiv A_p^2/R_{fp}$$

In the above equation, F_{MR} is the field-dependent damping force due to yield stress of the MR fluid. It can be expressed by

$$F_{MR} = A_p 2\frac{L}{h}\alpha H^\beta \tag{6.24}$$

The dynamic performance of the proposed damper is evaluated by computer simulation based on the dynamic model in Equation (6.23). Figure 6.15 presents the damping force simulated under harmonic excitation. Three different magnitudes of input displacement are considered, such as ±20 mm, ±50 mm, and ±100 mm. The results are demonstrated for various magnetic fields in frequency and time domains. It is observed that the damping force increases as the intensity of the applied magnetic field increases.

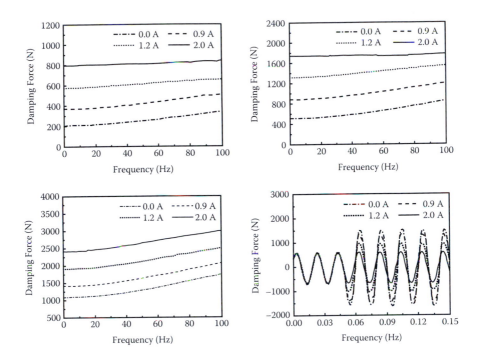

FIGURE 6.15
Dynamic characteristics of the MR impact damper. (From Woo, D. et al., *Journal of Intelligent Material Systems and Structures*, 18, 12, 2007. With permission.)

6.3.2 Collision Mitigation

A vehicle system is simplified to chassis, engine room, and bumper. Its motion is assumed to be in one direction. Figure 6.16 shows the vehicle model with the MR impact damper that has 3 degrees-of-freedom [28, 29]. The vehicle model includes an occupant model that has 2 degrees-of-freedom considering the upper and lower part of the body [30]. Therefore, the dynamic model of the vehicle system including occupant is derived as:

$$M_1\ddot{x}_1 = -2k_1(x_1 - x_2) + 2F_{cb}$$

$$M_2\ddot{x}_2 = 2k_1(x_1 - x_2) - 2k_2(x_2 - x_3)$$

$$M_3\ddot{x}_3 = 2k_2(x_2 - x_3) - 2k_3(x_3 - x_4 - R\sin\theta) \qquad (6.25)$$

$$M_4\ddot{x}_4 = k_3(x_3 - x_4 - R\sin\theta) - c_2(\dot{x}_4 - \dot{x}_3) - M_5 L\ddot{\theta}\cos\theta$$

$$M_5 L^2\ddot{\theta} = k_3(x_3 - x_4 - R\sin\theta)R\cos\theta + M_5 gL\sin\theta$$

where M_1 is the mass of the bumper, M_2 is the mass of engine room, M_3 is the mass of the frame, M_4 is the lower part of the body, and M_5 is the head mass.

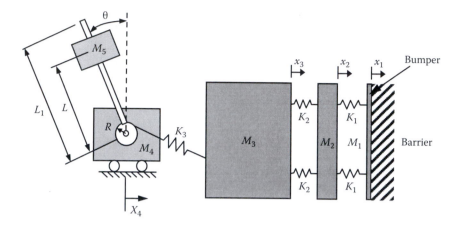

FIGURE 6.16
Mathematical model of the vehicle system.

In this section, the collision mitigation performance is evaluated by computer simulation. Initial conditions of the vehicle are imposed as:

$$\theta(0) = 0$$

$$x_1(0) = x_2(0) = x_3(0) = x_4(0) = 0 \tag{6.26}$$

$$\dot{x}_1(0) = \dot{x}_2(0) = \dot{x}_3(0) = \dot{x}_4(0) = v_0$$

During a vehicle crash, the collision process has three phases—crash initiation, airbag deployment, and occupant contact. Each phase requires different magnitudes of damping force. Therefore, a control algorithm for the MR damper is formulated according to crash time as:

$$F_{cb} = \begin{cases} 45\ kN & \text{if } 0 < t \le 30 \quad (m\,\text{sec}) \\ 15\ kN & \text{if } 30 < t \le 70 \quad (m\,\text{sec}) \\ 30\ kN & \text{if } t > 70 \quad\quad (m\,\text{sec}) \end{cases} \tag{6.27}$$

For performance evaluation, a vehicle crash severity index (VCSI) is introduced as a performance criterion. VCSI is calculated by deceleration data as:

$$VCSI = \frac{1}{T} \int_0^T (-\ddot{x})^2\, dt \tag{6.28}$$

Figure 6.17 presents collision mitigation performance at three different vehicle speeds such as 32 km/h, 48 km/h, and 56 km/h. In the case of crash speed of 48 km/h, occupant deceleration of the conventional vehicle

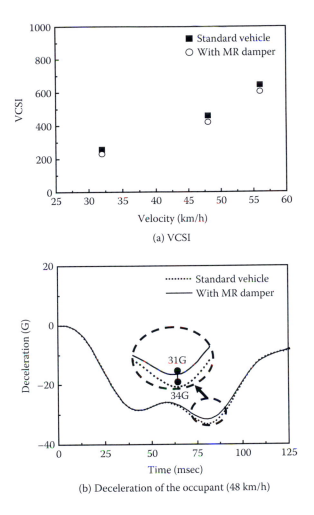

FIGURE 6.17

Comparison of the vehicle crash severity index. (From Woo, D. et al., *Journal of Intelligent Material Systems and Structures*, 18, 12, 2007. With permission.)

is 34 G. It decreases to 31 G for the vehicle with the MR impact damper installed. In the meantime, the VCSI value decreases from 459 to 421. Figures 6.18(a) and (b) present deformation of the vehicle body. At crash speed of 56 km/h, the deformation amount decreased from 65 cm to 58 cm by adopting the MR impact damper. From the results of VCSI and frame deformation, it is clearly observed that vehicle collision is successfully mitigated by imposing different damping forces for each crash phase. In addition, energy dissipation during the crash is evaluated in Figure 6.19. The area below the solid line is the energy dissipated by the MR damper. It is obvious from the results that the impact energy is well dissipated by activating the MR damper.

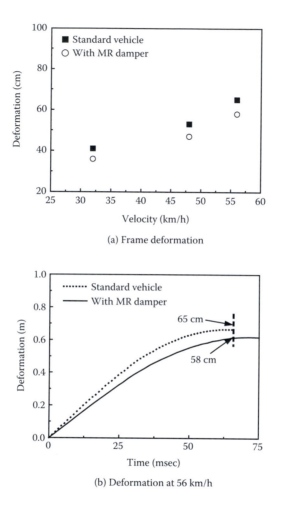

(a) Frame deformation

(b) Deformation at 56 km/h

FIGURE 6.18
Response comparison with and without MR damper. (From Woo, D. et al., *Journal of Intelligent Material Systems and Structures*, 18, 12, 2007. With permission.)

6.4 Some Final Thoughts

In this chapter, two interesting MR applications were discussed in the fields of vibration attenuation and collision mitigation of vehicles. In Section 6.2, a semi-active engine mount was considered for vibration attenuation and its vibration control performances were evaluated by considering a full-vehicle model. In the vehicles, the engine is a significant source of vibration, but a conventional passive mount has performance limitations. In order to resolve this problem, a semi-active mount utilizing MR fluids was introduced for vehicle engines. Under HILS configuration consisting of the full-vehicle

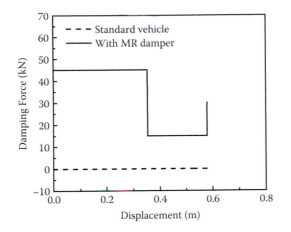

FIGURE 6.19
Energy dissipation with and without MR damper. (From Woo, D. et al., *Journal of Intelligent Material Systems and Structures*, 18, 12, 2007. With permission.)

model and the manufactured MR mount, both displacement and acceleration at engine C.G. point are favorably suppressed by activating the three MR engine mounts. In addition, it was shown that the acceleration level at the driver's position could be considerably reduced by employing the control input current to the MR engine mounts.

Section 6.3 introduced an impact damper available for vehicle bumpers to reduce driver injury and damage of the vehicle body. Collision mitigation is a very important research field in vehicle engineering. MR fluid is a potential candidate of this field. A vehicle collision process consists of three phases requiring different damping force levels: crash initiation, airbag deployment, and occupant contact. In this section, an analytical crash model of vehicles was derived under consideration of MR impact damper, occupants, and vehicle body structure. It has been simulated that VCSI and frame deformation could be effectively reduced by employing the MR impact damper with a simple step-wise control strategy according to the collision phase.

References

[1] Barber, D. E. and Carlson, J. D. 2010. Performance characteristics of prototype MR engine mounts containing glycol MR fluids. *Journal of Intelligent Material Systems and Structures* 21: 1509–1516.

[2] Mansour, H., Arzanpour, S., Golnaraghi, M. F., and Parameswaran, A. M. 2010. Semi-active engine mount design using auxiliary magnetorheological fluid compliance chamber. *Vehicle System Dynamics* 49: 449–462.

[3] Duclos, T. G. 1987. An externally tunable hydraulic mount which uses electro-rheological fluid. *SAE Technical Paper Series* 870963.

[4] Morishita, S. and Mitsui, J. 1992. An electronically controlled engine mount using electro-rheological fluid. *SAE Technical Paper Series* 922290.

[5] Williams, E. W., Rigby, S. G., Sproston, J. L., and Stanway, R. 1993. Electro-rheological fluids applied to an automotive engine mount. *Journal of Non-Newtonian Fluid Mechanics* 47: 221–238.

[6] Choi, S. B. and Choi, Y. T. 1999. Sliding mode control of a shear-mode type ER engine mount. *KSME International Journal* 13: 26–33.

[7] Choi, S. B. and Song, H. J. 2002. Vibration control of a passenger vehicle utilizing a semi-active ER engine mount. *Journal of Vehicle System Dynamics* 37: 193–216.

[8] Witteman, W. 1999. Improved vehicle crashworthiness design by control of the energy absorption for different collision situations. Ph.D. Dissertation, Eindhoven University of Technology, Automotive Engineering & Product Design Technology, Eindhoven, the Netherlands.

[9] Mizuno, K. and Kajzer, J., 1999. Compatibility problems in frontal, side, single car collisions and car-to-pedestrian accidents in Japan. *Accident Analysis Prevention* 31: 381–391.

[10] Witteman, W. 2005. Adaptive frontal structure design to achieve optimal deceleration pulses. *Proceedings of the 19th International Technical Conference on the Enhanced Safety of Vehicles*, Washington, D.C., pp. 1–8.

[11] Wågström, L., Thomson, R., and Pipkorn, B. 2004. Structural adaptivity for acceleration level reduction in passenger car frontal collisions. *International Journal of Crash* 9: 121–127.

[12] Jawad, S. A. W. 1996. Intelligent hydraulic bumper for frontal collision mitigation. *Crashworthiness and Occupant Protection in Transportation Systems ASME* 218: 181–189.

[13] Lee, D. Y., Choi, Y. T., and Wereley, N. M. 2002. Performance analysis of ER/MR impact damper systems using Herschel-Bulkley model. *Journal of Intelligent Material Systems and Structures* 13: 525–531.

[14] Ahmadian, M., Appleton R., and Norris, J. A. 2002. An analytical study of fire out of battery using magnetorheological dampers. *Shock and Vibration* 9: 129–142.

[15] Lee, H. S. and Choi, S. B. 2000. Control and response characteristics of a magneto-rheological fluid damper for passenger vehicles. *Journal of Intelligent Material Systems and Structures* 11: 80–87.

[16] Lee, H. S., Choi, S. B., and Lee, S. K. 2001. Vibration control of a passenger vehicle featuring MR suspension units. *Transactions of the Korean Society for Noise and Vibration Engineering* 11: 41–48.

[17] Song, H. J., Choi, S. B., Kim, J. H., and Kim, K. S. 2004. Performance evaluation of ER shock damper subjected to impulse excitation. *Journal of Intelligent Material Systems and Structures* 13: 625–628.

[18] Hong, S. R. and Choi, S. B. 2012. Vibration control of a structural system using magneto-rheological fluid mount. *Journal of Intelligent Material Systems and Structures* 16: 931–936.

[19] Duncan, M. R., Wassgren, C. R., and Krousgrill, C. M. 2005. The damping performance of a single particle impact damper. *Journal of Sound and Vibration* 286: 123–144.

[20] Lee, H. H. 2002. Vibration control of an engine mount featuring MR fluid. M.S. Thesis, Inha University, Incheon, South Korea.

[21] Lee, D. Y. and Wereley, N. M. 2000. Analysis of electro- and magneto-rheological flow mode dampers using Herschel-Bulkley model. *Proceedings of SPIE* 3989: 244–255.

[22] Watten, J. 1989. *Fluid Power Systems*, New York: Prentice Hall.

[23] Singh, R., Kim, G., and Ravinda, R. V. 1992. Linear analysis of automotive hydro-mechanical mount with emphasis on decoupler characteristics. *Journal of Sound and Vibration* 158: 219–243.

[24] Rao, S. S. 1995. *Mechanical Vibration*, New York: Addison-Wesley.

[25] Ooh, J. W. 1997. Optimization of nonlinear engine mount for the vibration reduction of a passenger car. Ph.D. Dissertaion, Inha University, Incheon, South Korea.

[26] Karnopp, D., Crosby, M. J., and Harwood, R. A. 1974. Vibration control using semi-active force generator. *ASME Journal of Engineering for Industry* 96: 619–626.

[27] Leitmann, G. 1994. Semi-active control for vibration attenuation. *Journal of Intelligent Material Systems and Structures* 5: 841–846.

[28] Elmarakbi, A. M. and Zu, J. W. 2004. Dynamic modeling and analysis of vehicle smart structures for frontal collision improvement. *International Journal of Automotive Technology* 5: 244–255.

[29] Kamal, M. 1979. Vibration analysis and simulation of vehicle-to-barrier impact. *SAE Transactions* 700414.

[30] Weaver, J. R. 1968. A simple occupant dynamics model. *Journal of Biomechanics* 1: 185–191.

7

Magnetorheological (MR) Brake System

7.1 Introduction

As discussed in earlier chapters, MR dampers and mounts are most actively and vibrantly researched in automotive engineering. Vehicle brake systems comprise another possible application of MR fluids. A brake is a mechanical device that inhibits motion. Brakes are generally applied to rotating wheels (or axles) to stop a vehicle. Most common brakes use friction to convert kinetic energy into heat, although several other methods of energy conversion may be employed. There are currently some efforts underway to develop a variable resistance brake utilizing MR fluids. Webb developed an exercise apparatus using an MR brake [1]. Avraam et al. proposed an MR brake for a wrist rehabilitation application [2]. Development devices associated with an MR brake to replace conventional systems, which usually possess a complex actuating mechanism, also recently received great consideration in the field of haptics [3–5]. Several applications in the automotive industry using MR brakes have also been proposed [4, 6].

MR brake analysis and optimal design have also been considered to improve performance. Karakoc, Park, and Suleman [7] and Li and Du [8] suggested several design criteria such as material selection, sealing, mixing scheme, and MR fluid selection to increase performance. Moreover, optimal design schemes for MR brakes with various objectives have been investigated to obtain optimal performance under specific constraints. In the work of Gudmundsson, Jonsdottir, and Thorsteinsson [9], a multi-objective optimization problem was conducted to aid the development of an MR brake in a prosthetic knee joint. To cover the effect of the heat resulting from vehicle cruising, Nguyen and Choi [10] undertook a geometric optimal design of an MR brake considering the zero-field friction heat. In the work of Park, Luz, and Suleman [6], via a multidisciplinary optimal design, a conclusion was given that, compared to an equivalent hydraulic brake, the mass of the replacing MR brake can be reduced significantly. In the literature, most of the optimal design work used the finite element method (FEM) in calculating the magnetic parameters. Indeed, with FEM, magnetic field strength (or magnetic flux density) can be obtained at any location in the device. Moreover, the accuracy can be improved by increasing the number of fundamental elements. It is remarked that the meshing job of the magnetic model is too complicated to do on your own. Thus, in the works cited previously, finite element analysis of the magnetic circuit is usually adopted with the aid of commercial software. However, the time consumption and commercial software cost issues cannot be ignored. In addition, in some particular models, it is not necessary to explore the value of every component. For example, in MR brakes, we only need to know the magnetic field strength (of the magnetic flux density) in MR fluids, from which the expression for braking torque is conducted. Therefore, a simpler numerical approach that requires less time to analyze the magnetic field of the desired components is necessary.

Effort was recently made to analytically model the magnetic circuit of MR brakes using an equivalent electric circuit [2] in which it was supposed that the flux is uniformly distributed over a cross-section of the components. Moreover, the relationship between field strength and flux density is assumed to be linear as well. Subsequently, a simple magnetic circuit method with an electrical analogy applying Kirchhoff's laws has been developed. Obviously, with many assumptions, this approach is expected to be inaccurate. Hence, it is useful to demonstrate the magnetic phenomena by solving engineering problems such as optimization that require high precision.

Consequently, the focus of Section 7.2 is the application of a magnetic circuit model associated with MR devices to the optimal design of a bi-directional MR (BMR) brake [11]. The BMR brake proposed in this work is a novel type whose operating principle differs significantly from conventional MR brakes. The BMR brake model consists of two components—a mechanical part and a magnetic circuit. While the mechanical part is modeled using Bingham's equation, the magnetic circuit is formulated by exploiting the proposed method. In this method, a special form of the MR fluid's cross-section in the brake is adopted in which the gap is much less than the length; hence, the variation of the magnetic parameters along this gap is insignificant. Therefore, it is possible to approximate that their values are constant along the gap of the MR fluid's cross-section since a new approach to determine the magnetic field and density of the MR fluid component is proposed. After that, an optimal design aiming to minimize mass is undertaken to assess the proposed method's accuracy and effectiveness. Because the design is expected to possess numerous constraints and potential local optima, a particle swarm optimization algorithm in combination with a gradient-based repair method is used. The optimal solution for this problem is then investigated and compared with that using the FEM with the aid of commercial software. Finally, an experiment of the manufactured BMR brake with the optimal parameters is undertaken to validate the accuracy of the proposed analysis methodology.

Section 7.3 discusses torsional vibration, an important issue in many mechanical systems with rotating mechanical components [12]. It is often a concern in power transmission systems using rotating shafts or couplings in which it can

cause failure if not well controlled. In power transmission systems, the generated torque (e.g., internal combustion engines of vehicles) may not be smooth,
or the component being driven may not react smoothly to the torque (e.g., recip-
rocating compressors). The components transmitting the torque can also generate non-smooth or alternating torques (e.g., worn gears, misaligned shafts).
Because the components in power transmission systems are not infinitely stiff,
irregular torque can cause vibration around the axes of rotation.

There have been a number of studies in the field of torsional vibration
control [13–18]. One well-known tool for reducing torsional vibration, the
dry friction damper, also known as a Lanchester damper, was addressed
in a very early paper by Den Hartog and Ormondroyd [18]. These authors
presented an analytical and experimental work performed with the goal of
determining the appropriate dry friction in a Lanchester damper to optimally damp the torsional vibration of a primary system. An important result
of this work is the equation relating the constant optimal friction torque to
the excitation torque acting on the primary system.

Based on this theory background, Ye and Williams [19] developed an exact
form of steady-state solution for a rotational friction damper, and a numerical
method was used to determine the optimum friction torque. Ye and Williams
[20, 21] also recently investigated the use of an MR fluid brake to control
torsional vibrations. In these studies, a configuration of torsional vibration
control featuring an MR brake was proposed, and two different strategies
were examined to control MR braking torque. In the first strategy, a constant
current is applied to the MR brake coil and the brake is implemented as a
passive friction damper with a variable friction torque. The second approach
involves implementation of the MR brake with a modified skyhook damping control algorithm in which the MR brake torque is adjusted according
to a comparison between the sign of absolute velocity of the primary system and the sign of the relative velocity between the MR brake and the primary system to add effective damping. Experimental results showed that
torsional vibration was significantly suppressed, and the MR brake was
readily applied to torsional vibration problems traditionally handled using
a Lanchester damper. However, optimal design of the MR brake was not
considered and a high cost-control system may have resulted because both
the absolute velocity of the primary system and the velocity of the MR brake
disc are required. More recently, Sun and Thomas [22] examined the control
of torsional rotor vibrations using an electrorheological (ER) fluid dynamic
absorber. Although a promised result was seen from the research, high electric field and low friction torque were always inherent limitations for implementing ER brakes in practical applications.

As discussed previously, torsional vibration control is very significant in
engineering, and the MR brake absorber is one potential candidate. Although
there has been much research on the application of the MR brake absorber
in the control of torsional vibration, its optimal design was not mentioned in
these studies. In addition, in earlier studies, the optimal friction torque was

determined at the resonance of the shaft system and depended only on the applied torque magnitude. However, it is obvious that the optimal friction torque also depends on the applied torque frequency.

Consequently, Section 7.3 focuses on two issues: (1) Determining the optimal friction torque of the friction absorber as a function of the applied torque frequency, and (2) optimally designing an MR brake absorber based on the optimal friction torque. To achieve these goals, a configuration of torsional vibration control of a rotating shaft using an MR brake absorber is first proposed and the braking torque of the MR brake is derived based on the Bingham plastic model of the MR fluid. Optimal design of the MR brake absorber is then investigated and a procedure to solve the optimal problem is proposed. The optimal torque for controlling torsional vibration is also examined. Based on the proposed optimal design procedure, the optimal design of a specific rotating shaft system is performed. The vibration control performances of rotating shaft systems employing the optimized MR brake absorber are then investigated.

7.2 Bi-directional MR Brake

7.2.1 Configuration and Torque Modeling

The configuration of the proposed BMR brake is demonstrated in Figure 7.1. It consists of two coils, two rotors, one outer casing, and MR fluid filling the gap between the rotors and the casing. In order to avoid two magnetic fields interfering with each other, a non-magnetic partition is inserted at the

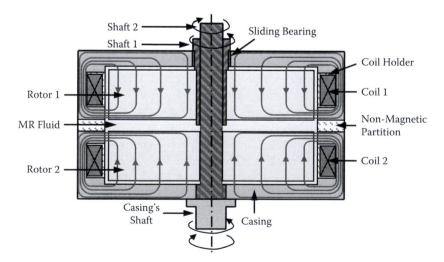

FIGURE 7.1
Configuration of the proposed BMR brake.

middle location of the casing. Two powers to supply to the coils are distinct such that the current magnitudes of these coils can be controlled independently. Unlike other (conventional) MR brakes, which only have one rotor and one stator (casing), the casing in this brake is not stationary. Indeed, it is fixed to a driving shaft, which might be connected to a one-dimensional handle in haptic applications. Moreover, two rotors are fixed to their respective shafts, which are transmitted from a driving bi-output source so that they rotate counter to each other. This assures that there exist two relative shear motions between the surfaces of two rotors and the outer casing even when the casing is at a stop.

As the current sources are applied to the coils, magnetic fields are generated in two separate zones as shown in Figure 7.1. Consequently, the solidification of the MR fluid at the gap between the rotors and outer casing occurs promptly. The shear friction between the casing and the rotors provides the resultant torque. The function of this torque can be either resistive or repulsive due to the schemes of applying the current sources to the coils as well as the rotation of the casing. It is assumed that the outer casing is motionless while two rotors rotate counter to each other. When only coil 1 (or 2) is excited, due to the solidification of MR fluid between the surfaces of rotor 1 (or 2) and casing, the outer casing tends to be pulled to rotate along with rotor 1 (or 2). The braking torque is the necessary torque to keep the casing still stationary and possesses the opposite direction with rotor 1 (or 2). In summary, the direction of the torque can be changed according to the excitation scheme of coil 1 or 2. Moreover, in the case of the casing rotating in the same direction as rotor 1, if only coil 1 is excited, the torque resulted from the yield stress in the MR fluid between the casing and rotor 1 is repulsive. In other words, the BMR brake works as a clutch. Otherwise, if only coil 2 is excited while the casing and rotor 1 rotate in the same direction, the generated torque is resistive and the brake works as a pure brake.

As seen from Figure 7.1, the total torque of the BMR brake is generated from three sources: the friction between the end-faces of the rotors and the casing; the friction between the annular faces of the rotors and the casing; and the dry friction resulted from the sealing scheme. The torque induced from these sources can be expressed in detail as:

$$T = \left| \overrightarrow{T_1} - \overrightarrow{T_2} \right| \tag{7.1}$$

where T_1 and T_2 are the induced torques contributed from the rotors 1 and 2, respectively, whose expressions are given by

$$T_1 = T_{ai} + T_{ei} + T_{fi}, \quad i = 1,2 \tag{7.2}$$

where T_{fi} is the torque due to dry friction between the surfaces of the rotor's shafts and the casing's shaft and from the sealing scheme, which can be determined via experiment; T_{ai} and T_{ei} are the induced torques transmitted

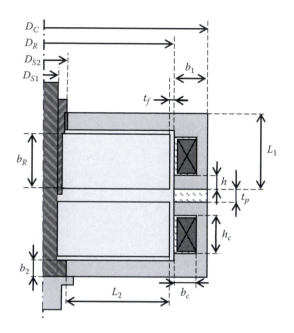

FIGURE 7.2
Significant geometric dimensions of a BMR brake.

from the friction between the MR fluid at the surfaces of rotors 1 and 2 and the faces of the casing, respectively, whose magnitude depends on the property of MR fluid and the applied field. In principle, by applying appropriate current sources to the coils, the total induced torque can be eliminated completely. With the geometric parameters shown in Figure 7.2, the expressions for field-dependent torques T_{ai} and T_{ei} can be given in the following forms:

$$T_{ai} = 2\pi \left(\frac{D_R}{2} \right)^2 \int_0^{b_R} \tau_{ai}\, dz, \quad i = 1,2 \tag{7.3}$$

$$T_{ei} = 2\pi \int_{\frac{D_{S2}}{2}}^{\frac{D_R}{2}} r^2 \tau_{ei}\, dr, \quad i = 1,2 \tag{7.4}$$

where, τ_{ai} and τ_{ei} are the shear stresses acting on the MR fluid at the surfaces of rotors i ($i = 1,2$) and the faces of the casing, respectively, whose values can be mathematically expressed by Bingham's model as:

$$\tau_{ai} = \tau_{yai} + K\dot{\gamma}_{ai} \tag{7.5}$$

$$\tau_{ei} = \tau_{yei} + K\dot{\gamma}_{ei} \tag{7.6}$$

where K is called the consistency; τ_{yai} and τ_{yei} are the yield stresses of MR fluid at the surfaces of rotors i ($i = 1,2$) and casing, respectively. The variation of these yield stresses depends on the property of MR fluid, which is available from the manufacturer's datasheet, and the magnitude of the applied currents to the coils. In Equation (7.5) and Equation (7.6), $\dot{\gamma}_{ai}$ and $\dot{\gamma}_{ei}$ are the shear rate of MR fluid at the gap between the annular faces and the end faces whose values can be determined as:

$$\dot{\gamma}_{ai} = \frac{D_R\left|\overrightarrow{\Omega_i} - \overrightarrow{\Omega_c}\right|}{2t_f}, \quad i = 1,2 \tag{7.7}$$

$$\dot{\gamma}_{ei} = \frac{r\left|\overrightarrow{\Omega_i} - \overrightarrow{\Omega_c}\right|}{t_f}, \quad i = 1,2 \tag{7.8}$$

where Ω_i and Ω_c are the angular velocities of the rotors and casing, respectively. By substituting Equation (7.5) to Equation (7.8) in Equation (7.3) and (7.4), the field-dependent torques can be expressed as:

$$T_{ai} = 2\pi\left(\frac{D_R}{2}\right)^2 \int_0^{b_R} \tau_{yai}\, dz + \frac{\pi K D_R^3 b_R \left|\overrightarrow{\Omega_i} - \overrightarrow{\Omega_c}\right|}{4t_f}, \quad i = 1,2 \tag{7.9}$$

$$T_{ei} = 2\pi \int_{\frac{D_{S2}}{2}}^{\frac{D_R}{2}} r^2 \tau_{yei}\, dr + \frac{\pi K\left[\left(\frac{D_R}{2}\right)^4 - \left(\frac{D_{S2}}{2}\right)^4\right]\left|\overrightarrow{\Omega_i} - \overrightarrow{\Omega_c}\right|}{2t_f}, \quad i = 1,2 \tag{7.10}$$

The values of the yield stresses τ_{yai} and τ_{yei} do not vary significantly in the shear surfaces. Therefore, for simplicity, these values can be considered as constants. Consequently, Equation (7.9) and Equation (7.10) can be rewritten in simpler forms as:

$$T_{ai} = \frac{\pi D_R^2 b_R \tau_{yai}}{2} + \frac{\pi K D_R^3 b_R \left|\overrightarrow{\Omega_i} - \overrightarrow{\Omega_c}\right|}{4t_f}, \quad i = 1,2 \tag{7.11}$$

$$T_{ei} = \frac{\pi\left(D_R^3 - D_{S2}^3\right)\tau_{yei}}{12} + \frac{\pi K\left(D_R^4 - D_{S2}^4\right)\left|\overrightarrow{\Omega_i} - \overrightarrow{\Omega_c}\right|}{32t_f}, \quad i = 1,2 \tag{7.12}$$

It is noteworthy that the torque induced by the viscosity and dry friction is insignificant compared to that induced by the magnetic field. Therefore, the second terms in the right-hand side of the equations as well as dry friction torques can be neglected in the optimal design process.

7.2.2 Magnetic Circuit

As aforementioned, the magnitude of the yield stresses of MR fluid at the different positions in the BMR brake depends on the magnetic field applied to MR fluid. In magnetic field theory, there are two fundamental laws of magnetic field and flux: Ampere's and Gauss's laws. Ampere circuital law states that the line integral of a magnetic field about any closed path is equal to the direct current enclosed by that path. In other words, Ampere's law can be expressed by the following equation:

$$\oint_C \vec{H} \cdot \vec{dl} = I_{nc} \tag{7.13}$$

where I_{nc} is the total net current that penetrates through the surface that is enclosed by the curve C.

Gauss's law states that for each volume in space, the magnetic flux exiting the volume is exactly the same as the one entering the volume. This law can be expressed in a popular form as:

$$\oint_S \vec{B} \cdot \vec{dA} = 0 \tag{7.14}$$

where B is the magnetic density and S is the closed surface that encloses the volume. Solving Equation (7.13) and Equation (7.14) is generally sophisticated because of the complexity of the surface in consideration as well as the variation of magnetic parameters in devices. However, it is remarkable that in an MR brake, all we need is the magnetic parameters in the MR fluid; from that, the braking torque can be computed. Another considerable property is the special shape of the cross-section MR fluid in which the thickness is much less than the length. By exploiting these properties, a new magnetic circuit analysis is proposed as follows. Figure 7.3 shows three possible configurations of magnetic circuit of a quarter of the BMR brake where the positions of the coils are different. In configuration (a), the coils locate partly in the end portion of the casing whereas they locate completely in the radial and end portions of the casing in configuration (b) and (c), respectively. It is noteworthy that in configuration (c), most flux flows in the casing instead of penetrating through the MR fluid elements. Consequently, the induced torque is very low in this case. In other words, in an engineering optimization problem, the optimal solution does not exist with configuration (c). Therefore, in the optimization problem, this case can be rejected from consideration by implementing the following constraint:

$$h \le b_R + t_f \tag{7.15}$$

where h is the height of the coil, and b_R and t_f are the thickness of the rotor and MR fluid elements, respectively.

In the case of configuration (a), the volume is discretized into 11 elements as shown in the figure, whereas elements 8 and 11 are MR fluid and the

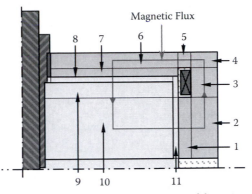

(a) Coils locate partly in the end portions of the casing

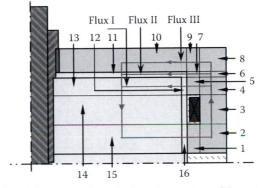

(b) Coils locate completely in the radial portions of the casing

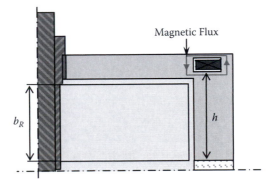

(c) Coils locate completely in the end portions of the casing

FIGURE 7.3
Configurations of the magnetic circuit of a BMR brake.

others are magnetic materials. As shown in the figure, the magnetic flux is considered as a closed line that penetrates at the middle positions of the cross-sections of all elements. Therefore, Equation (7.14) can be restated as:

$$B_1 A_1 = B_2 A_2 = \ldots = B_{11} A_{11} = \Phi \tag{7.16}$$

where Φ is the magnetic flux penetrating through the mean cross-sections $A_1, A_2, \cdots, A_{11}$ of the elements. The mean cross-section of an element is obtained as the average of the entering and exiting cross-sections of that element. The expressions for the elements in configuration (a) of the BMR brake are computed as:

$$A_i = \frac{A_{INi} + A_{OUTi}}{2}, \quad i = 1 \ldots 11 \tag{7.17}$$

where

$$A_{IN1} = \pi D_R h$$

$$A_{OUT1} = A_{IN2} = \pi(D_R + 2b_c)h$$

$$A_{OUT2} = A_{IN3} = A_{OUT3} = A_{IN4} = \pi\left[\frac{D_c^2}{4} - \frac{(D_R + 2b_c)^2}{4}\right]$$

$$A_{OUT4} = A_{IN5} = \pi(D_R + 2b_c)(L_1 - h - h_c)$$

$$A_{OUT5} = A_{IN6} = \pi D_R (L_1 - h - h_c)$$

$$A_{OUT6} = A_{IN7} = \pi\left(\frac{D_R^2}{4} - \frac{D_{s2}^2}{4}\right)$$

$$A_{OUT7} = A_{IN8} = A_{OUT8} = A_{IN9} = A_{OUT9} = A_{IN10} = \pi\left[\frac{(D_R - 2t_f)^2}{4} - \frac{D_{s2}^2}{4}\right]$$

$$A_{OUT10} = A_{IN11} = \pi(D_R - 2t_f)h$$

$$A_{OUT11} = A_{IN1} = \pi D_R h$$

$B_1, B_2, \cdots, B_{11}$ are the magnetic densities on the mean cross-sections of the elements. They have the highly nonlinear relation with the magnetic field depending on materials, which is available in magnetic materials handbooks. The relations can be expressed symbolically for steel and MR fluid materials as:

$$H = f_s(B) \quad or \quad B = g_s(H) \qquad \text{for steel} \tag{7.18}$$

$$H = f_{MR}(B) \quad or \quad B = g_{MR}(H) \qquad \text{for MR fluid} \tag{7.19}$$

In addition, from Figure 7.3(a), by choosing the closed path to be coincident with the flux line, Equation (7.13) can be reformed as:

$$\sum_{k=1}^{11} H_k l_k = I_{nc} \tag{7.20}$$

where l_k is the length of the element k given by

$$l_1 = b_c$$

$$l_2 = \frac{b_1 - b_c}{2} + \frac{h}{2}$$

$$l_3 = h_c$$

$$l_4 = \frac{L_1 - h_c - h}{2} + \frac{b_1 - b_c}{2}$$

$$l_5 = b_c$$

$$l_6 = \frac{L_2 + t_f}{2} + \frac{L_1 - h_c - h}{2} \tag{7.21}$$

$$l_7 = h_6 + h - (L_1 - b_2)$$

$$l_8 = t_f$$

$$l_9 = L_1 - h - b_2 - t_f$$

$$l_{10} = \frac{h}{2} + \frac{L_2}{2}$$

$$l_{11} = t_f$$

By substituting Equation (7.16) through Equation (7.19) into Equation (7.20), the magnetic density of the MR fluid elements (elements 8 and 11) can be obtained as:

$$\sum_{k=1,k\neq 8,11}^{11} f_s\left(\frac{A_8}{A_k}B_8\right)l_k + \sum_{k=8,11} f_{MR}\left(\frac{A_8}{A_k}B_8\right)l_k = I_{nc} \tag{7.22}$$

$$\sum_{k=1,k\neq 8,11}^{11} f_s\left(\frac{A_{11}}{A_k}B_{11}\right)l_k + \sum_{k=8,11} f_{MR}\left(\frac{A_{11}}{A_k}B_{11}\right)l_k = I_{nc} \tag{7.23}$$

Since $f_s(\ )$ and $f_{MR}(\ )$ are monotonic functions with respect to B, Equation (7.22) and Equation (7.23) can be solved easily using numerical method tools.

In the case of the magnetic circuit shown in Figure 7.3(b), the volume is discretized into 16 elements, whereas the 11, 12, and 16 are MR fluid elements and the others are magnetic material ones. As shown in the figure, the magnetic flux is split into three branches. Among them, flux III is expected to be insignificant because it penetrates through the MR fluid element, which has a high permeability, a much longer distance than the others do. Therefore, flux III can be neglected from consideration without affecting the result significantly. From Figure 7.3(b), Equation (7.14) is rewritten as:

$$\begin{cases} B_1 A_1 = B_2 A_2 = B_3 A_3 = B_4 A_4 = B_{13} A_{13} = B_{14} A_{14} = B_{15} A_{15} = B_{16} A_{16} = \Phi \\ B_{4I} A_{4I} = B_6 A_6 = B_8 A_8 = B_9 A_9 = B_{10} A_{10} = B_{11} A_{11} = B_{13I} A_{13I} = \Phi_I \\ B_{4II} A_{4II} = B_5 A_5 = B_{12} A_{12} = B_{13II} A_{13II} = \Phi_{II} \end{cases} \qquad (7.24)$$

$$\Phi = \Phi_I + \Phi_{II} \qquad (7.25)$$

In addition, from Figure 7.3(b), Equation (7.13) can be reformed as:

$$\begin{cases} H_1 l_1 + H_2 l_2 + H_3 l_3 + H_4 l_4 + H_{4I} l_{4I} + H_6 l_6 + H_8 l_8 + H_9 l_9 + H_{10} l_{10} \\ \qquad + H_{11} l_{11} + H_{13I} l_{13I} + H_{13} l_{13} + H_{14} l_{14} + H_{15} l_{15} + H_{16} l_{16} = I_{nc} \\ H_1 l_1 + H_2 l_2 + H_3 l_3 + H_4 l_4 + H_{4II} l_{4II} + H_5 l_5 + H_{12} l_{12} \\ \qquad + H_{13II} l_{13II} + H_{13} l_{13} + H_{14} l_{14} + H_{15} l_{15} + H_{16} l_{16} = I_{nc} \end{cases} \qquad (7.26)$$

The expressions for the length of the magnetic flux of the elements in configuration (b) of the BMR brake, which encloses the total net current, are given as:

$$l_1 = b_c$$

$$l_2 = \frac{b_1 - b_c}{2} + \frac{h}{2}$$

$$l_3 = h_c$$

$$l_{4I} = \frac{b_1 - b_c}{2}$$

$$l_{4II} = \frac{L_1 - b_2 - t_f - h_c - h}{2}$$

$$l_4 = \frac{L_1 - b_2 - t_f - h_c - h}{2}$$

$$l_5 = b_c$$

$$l_6 = t_f$$

$$l_7 = b_c$$

$$l_8 = \frac{b_2}{2} + \frac{b_1 - b_c}{3}$$

$$l_9 = b_c \tag{7.27}$$

$$l_{10} = \frac{L_2}{2} + t_f + \frac{b_2}{2}$$

$$l_{11} = l_{12} = t_f$$

$$l_{13I} = \frac{L_2}{2}$$

$$l_{13II} = \frac{L_1 - b_2 - t_f - h_c - h}{2}$$

$$l_{13} = \frac{L_1 - b_2 - t_f - h_c - h}{2}$$

$$l_{14} = h_c$$

$$l_{15} = \frac{h}{2} + \frac{L_2}{2}$$

$$l_{16} = t_f$$

The mean cross-sections of the elements in configuration (b) of the BMR brake are computed as:

$$A_i = \frac{A_{INi} + A_{OUTi}}{2}, \quad i = 1...11, i \neq 4,13$$

$$A_{4I} = \frac{A_{IN4} + A_{OUT4I}}{2}$$

$$A_{4II} = A_4 = A_{IN4} \tag{7.28}$$

$$A_{13I} = \frac{A_{IN13I} + A_{OUT13}}{2}$$

$$A_{13II} = A_{13} = A_{OUT13}$$

where

$$A_{IN1} = \pi D_R h$$

$$A_{OUT1} = A_{IN2} = \pi (D_R + 2b_c) h$$

$$A_{OUT2} = A_{IN3} = A_{OUT3} = A_{IN4} = \pi \left[\frac{D_c^2}{4} - \frac{(D_R + 2b_c)^2}{4} \right]$$

$$A_{OUT4I} = A_{IN5} = \pi (D_R + 2b_c)(L_1 - b_2 - t_f - h - h_c)$$

$$A_{OUT5} = A_{IN2} = \pi D_R (L_1 - b_2 - t_f - h - h_c)$$

$$A_{OUT12} = A_{IN13I} = \pi (D_R - 2t_f)(L_1 - b_2 - t_f - h - h_c)$$

$$A_{OUT4II} = A_{IN6} = A_{OUT6} = A_{IN8} = \pi \left[\frac{D_c^2}{4} - \frac{(D_R + 2b_c)^2}{4} \right]$$

$$A_{OUT8} = A_{IN9} = \pi (D_R + 2b_c) b_2$$

$$A_{OUT9} = A_{IN10} = \pi D_R b_2$$

$$A_{OUT10} = A_{IN11} = A_{OUT11} = A_{IN13II} = A_{OUT13} = A_{IN14} = A_{OUT14} = A_{IN15}$$

$$= \pi \left(\frac{D_R^2}{4} - \frac{D_s^2}{4} \right)$$

$$A_{OUT15} = A_{IN16} = \pi (D_R - 2b_c) h$$

$$A_{OUT16} = \pi D_R h$$

Similarly, Equation (7.18), Equation (7.19), as well as Equation (7.24) to Equation (7.26) compose a solution for magnetic density of the MR fluid elements (elements 11, 12, and 16).

7.2.3 Optimal Design

An optimal design problem with the objective to minimize the weight of the BMR brake is undertaken in this section. Several factors are significant in the design process such as weight, power, and braking torque. In our work, whereas the power is neglected because the BMR brake is expected to be applied in low-power applications, the weight and the braking torque are considered as important factors. In the optimization, the objective is to minimize the weight of the BMR brake while the braking torque generated from each rotor is expected not to be less than 2 Nm. Moreover, for the BMR brake to be developed successfully, several geometric constraints must be assured. Figure 7.2 demonstrates the configuration of a BMR brake with all significant geometric dimensions. As shown in this figure, the design variables (DVs) are the casing length L_1, the radial length of the rotors L_2, the radial and axial widths of the casing b_1, b_2, and the length, width, and height of the coil h_c, b_c, h. The other dimensions are fixed for feasibility of manufacture. For the coil not to be out of the scope of the casing, the width of the coil (b_c) should be less than the radial width of the casing (b_1) and the height of the coil should be smaller than the length of the casing (L_1). Moreover, in order to avoid configuration (c) as shown in Figure 7.3(c), the constraint (7.15) has to be included. In short, the optimization problem of a BMR brake is to minimize the weight of the BMR brake subject to:

$$Torque > 2Nm \quad \text{(for each rotor)}$$

$$h_c < b_1$$

$$h_c + h < L_1 \tag{7.29}$$

$$h \le b_R + t_f$$

$$\min \mathbf{DV}_i \le \mathbf{DV}_i \le \max \mathbf{DV}_i \quad i = 1, \cdots, 7$$

where

$$\min \mathbf{DV} = \begin{bmatrix} L_{1\min} & L_{2\min} & b_{1\min} & b_{2\min} & h_{c\min} & b_{c\min} & h_{\min} \end{bmatrix}$$

$$\max \mathbf{DV} = \begin{bmatrix} L_{1\max} & L_{2\max} & b_{1\max} & b_{2\max} & h_{c\max} & b_{c\max} & h_{\max} \end{bmatrix}$$

The optimization tool used in the study is the particle swarm optimization (PSO) algorithm, which is one of the newest heuristic algorithms in the literature. Moreover, like the genetic algorithm (GA), it is very effective in solving problems that have multiple local optima. This algorithm, which was first developed by Eberhart and Kennedy [23], is inspired from a process of looking for the food of a swarm of fishes or birds. Besides searching for the best by own, a particle receives the information about the global best from its swarm and tends to adjust the searching direction based on this global information. An outstanding advantage of PSO compared to GA, which is also a heuristic optimization, is the ease of programming. The procedure of PSO can be summarized in two steps. First, a swarm of particles (or solutions) and their velocity are generated randomly. Second, the particles are "flown" through searching space by updating their velocity and position by the two following equations [24]:

$$\mathbf{V}_i(k+1) = w \times \mathbf{V}_i(k) + c_1 \times r_1 \times \left[\mathbf{pDV}_i(k) - \mathbf{DV}_i(k) \right] + c_2 \times r_2 \times \left[gDV(k) - \mathbf{DV}_i(k) \right] \tag{7.30}$$

$$\mathbf{DV}_i(k+1) = \mathbf{DV}_i(k) + \mathbf{V}_i(k+1) \tag{7.31}$$

where $V_i(k)$ is the instant velocity of the particle DV_i. The subscript i stands for the ith particle in the swarm. c_1 and c_2 are the acceleration constants and are usually set to 2 [24]. r_1 and r_2 are two random numbers between 0 and 1, respectively. $pDV_i(k)$ and $gDV(k)$ are the best DVs that the ith particle and global swarm reach at the instance k, respectively. w is the inertial weight whose value affects the searching quality for optimization. The larger the value it is, the less possibility it is trapped in a local optimum. However, with the large w, the possibility of missing the true optimum is considerable. In general, it is usually assigned to be 0.8.

It is known that PSO is very effective in solving unconstraint optimization problems. However, in order to apply this algorithm to the current optimal design, which consists of numerous constraints, it is necessary to adopt a constraint-handling method. Thus far, penalty and repair methods are the most widely used approaches. In our work, the gradient-based repair method proposed in Reference [25], which is proven effective when integrated with GA in solving engineering optimization problems, is adopted for the optimal design problem. The procedure for the repair method is summarized as:

Step 1: Determine the derivatives of the constraint with respect to the DVs as:

$$\nabla_{DV} V = \begin{bmatrix} \nabla_{DV} g(DV) \\ \nabla_{DV} h(DV) \end{bmatrix} \tag{7.32}$$

where, $g()$ and $h()$ are the sets of inequality and equality constraints. Sometimes, in some applications, these derivatives are unlikely to be obtained analytically. Instead, the following approximate approach is used:

$$\nabla_{DV} V = \begin{bmatrix} \nabla_{DV} g(DV) \\ \nabla_{DV} h(DV) \end{bmatrix} = \frac{1}{e} \begin{bmatrix} g(DV|DV_i = DV_i + e) - g(DV) \\ h(DV|DV_i = DV_i + e) - h(DV) \end{bmatrix} \tag{7.33}$$

where e is a small augmentation.

Step 2: Compute the degree of constraint violation by the following equation:

$$\Delta V = \begin{bmatrix} \min\{0, -g(DV)\} \\ h(DV) \end{bmatrix} \tag{7.34}$$

Step 3: The change of solution ΔDV is obtained based on the following relationship:

$$\Delta V = \nabla_{DV} V \times \Delta DV \Rightarrow \Delta DV = (\nabla_{DV} V)^{-1} \times \Delta V \tag{7.35}$$

where $(\nabla_{DV} V)^{-1}$ is the Moore-Pensore inverse of $\nabla_{DV} V$.

Step 4: Update the solution vector by

$$DV(k+1) = DV(k+1) + \Delta DV \tag{7.36}$$

Step 5: Check for constraint violations of the DVs. If any constraint violations still exist, repeat Steps 1 through 4. This repair process should be made five times at most.

In short, the flowchart for optimal design of the BMR brake proposed in this work is given in detail in Figure 7.4.

7.2.4 Results and Discussions

In this section, the solution resulting from the optimization problem developed in the previous section is obtained and analyzed. The materials for the BMR brake and their magnetic properties are listed in Table 7.1. As shown in the table, the shafts of the rotors are made of copper alloy, which is a non-magnetic material. This prevents the magnetic flux from flowing to the insignificant positions and helps save power. The magnetic property of 1008 steel is demonstrated in Figure 7.5. By using the curve-fitting method, the relationship between the magnetic density and magnetic intensity are obtained in the following polynomial forms:

$$H = K_{s0}B^5 + K_{s1}B^4 + K_{s2}B^3 + K_{s3}B^2 + K_{s4}B + K_{s5} \tag{7.37}$$

where

$$K_s = \begin{cases} [\ 0\ \ 1.82\ -3.63\ \ 1.782\ \ 0.387\ \ 0] & B \le 1.5 \\ [-1419.52\ \ 13551.37\ -50744.31\ \ 93520.50\ -85032.46\ \ 30566.42] & B > 1.5 \end{cases}$$

The magnetic and yield stress properties of the MR fluid used, which is MRF-132-DG type from Lord Corporation [26], are also given in Figure 7.6 and Figure 7.7, respectively. As mentioned previously, their relationship can be obtained in the polynomials as:

$$H = K_{MR0}B^5 + K_{MR1}B^4 + K_{MR2}B^3 + K_{MR3}B^2 + K_{MR4}B + K_{MR5} \tag{7.38}$$

where

$$K_{MR} = \begin{bmatrix} -163.383 & 405.194 & -143.137 & 55.429 & 141.61 & 0 \end{bmatrix}$$

Alternatively, in the reverse:

$$B = P_{MR0}H^5 + P_{MR1}H^4 + P_{MR2}H^3 + P_{MR3}H^2 + P_{MR4}H + P_{MR5} \tag{7.39}$$

where

$$P_{MR} = \begin{bmatrix} 2.7 \times 10^{-14} & -6.2 \times 10^{-11} & 5.64 \times 10^{-8} & -2.6 \times 10^{-5} & 7.56 \times 10^{-3} & 0 \end{bmatrix}$$

$$\tau_y = m_{MR0}B^6 + m_{MR1}B^5 + m_{MR2}B^4 + m_{MR3}B^3 + m_{MR4}B^2 + m_{MR5}B + m_{MR6} \tag{7.40}$$

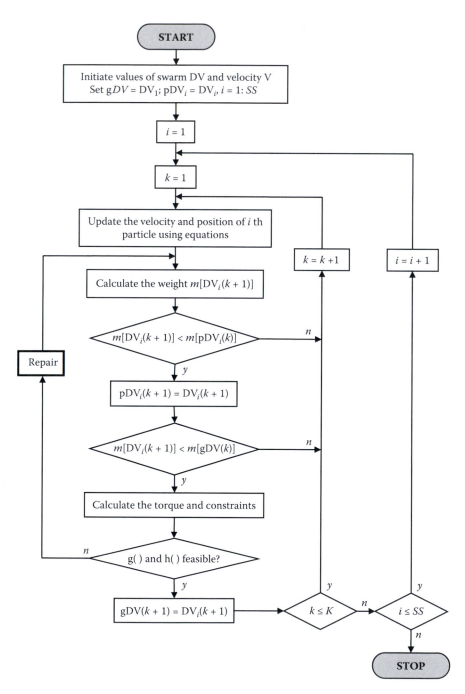

FIGURE 7.4
Flowchart for optimal design of a BMR brake using the proposed analysis method.

TABLE 7.1

Materials and Magnetic Properties of the Elements in the BMR Brake

Components	Materials	Magnetic Properties
Casing	1008 Steel	$B–H$ relationship as in Figure 7.5
Casing's shaft	1008 Steel	$B–H$ relationship as in Figure 7.5
Rotor 1	1008 Steel	$B–H$ relationship as in Figure 7.5
Rotor 2	1008 Steel	$B–H$ relationship as in Figure 7.5
Rotor 1's shaft	Copper	$B = 0$
Rotor 2's shaft	Copper	$B = 0$
Sliding bearing	Copper	$B = 0$
Coil 1	Copper	$B = 0$
Coil 2	Copper	$B = 0$
Coil holder	Aluminum	$B = 0$
Partition	Aluminum	$B = 0$

where

$$m_{MR} = \begin{bmatrix} 118.178 & -338.354 & 285.365 & -68.843 & 52.378 & 0 \end{bmatrix}$$

The gap of the MR fluid in the brake is fixed to be 1 mm. In addition, the other dimensions such as diameter of the rotor shafts and the thickness of the partition are preset based on their strength and manufacturing issues. The size of the coil wire is 24 gauge, with diameter approximately 0.5 mm. With this size, the maximum applied current that the wire can withstand

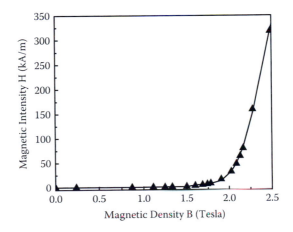

FIGURE 7.5

B–H curve of 1008 steel. (From Nguyen, P.B. et al., *Smart Materials and Structures,* 20, 12, 2011. With permission.)

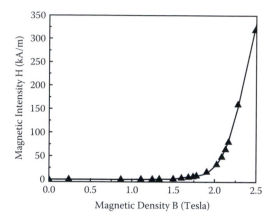

FIGURE 7.6
H–B curve of MR fluid. (From Nguyen, P.B. et al., *Smart Materials and Structures*, 20, 12, 2011. With permission.)

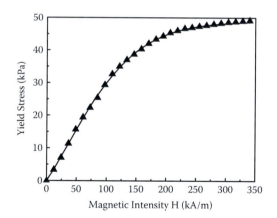

FIGURE 7.7
H–τ curve of MR fluid. (From Nguyen, P.B. et al., *Smart Materials and Structures*, 20, 12, 2011. With permission.)

is expected to be 1.5 A. Therefore, with a size 1×1 mm of the coil holder, a maximum of four turns of coil can be arranged. In other words, the current density is computed to be 6 A/mm^2. Therefore, with area of the cross-section $(b_c \times h_c)$ of a coil and the thickness ($t_c = 0.5$mm) of the holder, the value of the net current can be obtained from Figure 7.2 as:

$$I_{nc} = 6(b_c - t_c)(h_c - 2t_c) \tag{7.41}$$

Figure 7.8 shows the optimal solution of the BMR brake using the proposed magnetic analysis and PSO algorithm. It is noticed that the optimal solution occurs in configuration (a). The mean magnetic densities of MR fluid

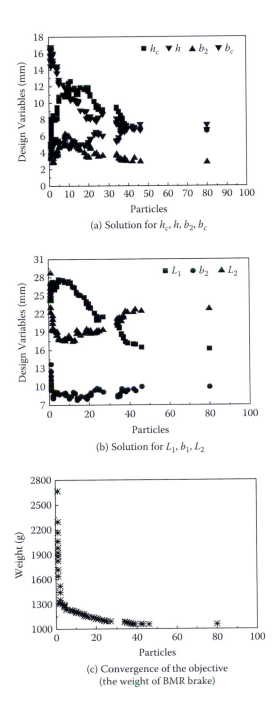

(a) Solution for h_c, h, b_2, b_c

(b) Solution for L_1, b_1, L_2

(c) Convergence of the objective
(the weight of BMR brake)

FIGURE 7.8
Solution of the optimal design using the proposed magnetic circuit analysis. (From Nguyen, P.B. et al., *Smart Materials and Structures*, 20, 12, 2011. With permission.)

TABLE 7.2

Optimal Values of the Parameters in the Optimization Using the Proposed
Analysis and FEA Methods

Parameters	Optimal Values		
	Proposed Analysis Method	FEA	Rounded for Manufacture
Casing length L_1 (mm)	16.202	15.582	16
Radial length of the rotors L_2 (mm)	22.778	23.389	23
Radial widths of the casing b_1 (mm)	9.925	9.925	10
Axial widths of the casing b_2 (mm)	2.824	2.794	3
Length of the coil h_c (mm)	6.692	6.334	6.5
Width of the coil b_c (mm)	7.364	7.521	7.5
Height of the coil h (mm)	6.685	6.454	6.5
Total net current I_{nc} (A)	234.42	224.70	231
Weight (g)	1055.6	1047.8	1065
Computation time	< 3 min	> 8 h	

at elements 8 and 11 are 0.3936 T and 0.6917 T, respectively. The torque value
for each rotor obtained from the optimization process is 2 Nm. The resultant
total net current applied to one coil of the BMR brake is 234.42 A. The optimal value of the DVs is given in Table 7.2. As shown in Figure 7.8, the swarm
size used in the optimization procedure is 100 and the objective nearly converges at the 50th particle. The effort of the others after this particle to find
the better solution is not very significant. It is noteworthy that because the
particles initially are chosen randomly, the convergent rate of the objective is
not the same for each time of process running. Nevertheless, the objective
always converges to a definite optimal value.

To assess the accuracy of the optimal design using the proposed magnetic
circuit analysis, the Maxwell software package from Ansoft Corporation [27]
is used to analyze the performance of the BMR brake. Figure 7.9 shows the
magnetic density of the BMR brake with the optimal DVs obtained previously.
It is shown that the magnetic densities at the MR fluid positions are approximately the same as those obtained from the optimization process. The torque
value obtained from the finite element analysis (FEA) is 2.08 Nm, which is
4% different in comparison with that obtained from the optimization process.
In addition, in order to investigate the geometric difference of the proposed
optimal design with that using FEA, an optimization procedure using FEA is
undertaken for comparison. Figure 7.10 shows the flowchart for this optimization procedure via FEA. As shown in Figure 7.10, the Maxwell FEA tool is used
instead of the proposed one. The optimal solution for this procedure is shown
in Figure 7.11 and Table 7.2. It is noted that there are no significant differences
in geometry between the two procedures using the proposed magnetic circuit
analysis and FEA. Therefore, it can be concluded that the proposed magnetic
analysis as well as the corresponding optimization procedure is effective and

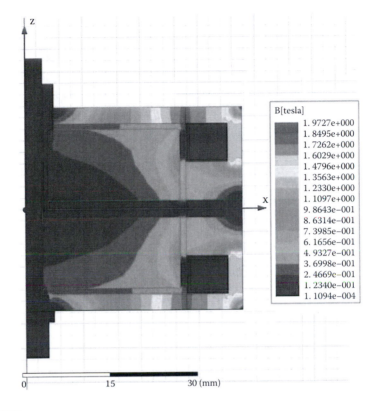

FIGURE 7.9
Magnetic density of the BMR brake with the obtained DVs. (From Nguyen, P.B. et al., *Smart Materials and Structures*, 20, 12, 2011. With permission.)

considerably accurate. Moreover, it is remarkable that the computation time of the optimal design with the proposed magnetic circuit analysis is drastically faster than that using FEA. In detail, the computation time when using the proposed magnetic circuit analysis is just about 3 min, while it takes over 8 h to solve the optimization problem using FEA.

In order to validate the accuracy of the optimal design with the proposed magnetic circuit analysis, an experiment with a realized BMR brake is undertaken. The geometric dimensions of this brake are obtained via the optimal design process of this work. For convenience in manufacturing, these dimensions are rounded with the step 0.5 mm. All significant dimensions for manufacture of the brake are given in Table 7.2. The manufactured components and assembly of the device are shown in Figure 7.12. In the experiment, torque value is obtained thanks to a torque sensor connected with the shaft of the casing. In addition, two rotor shafts are transmitted by a driving DC motor with revolutions of 30 rpm and a bi-output gearbox so that they can rotate counter to each other. In order to obtain the maximum braking torque in one direction, one coil is excited with maximum current value while the

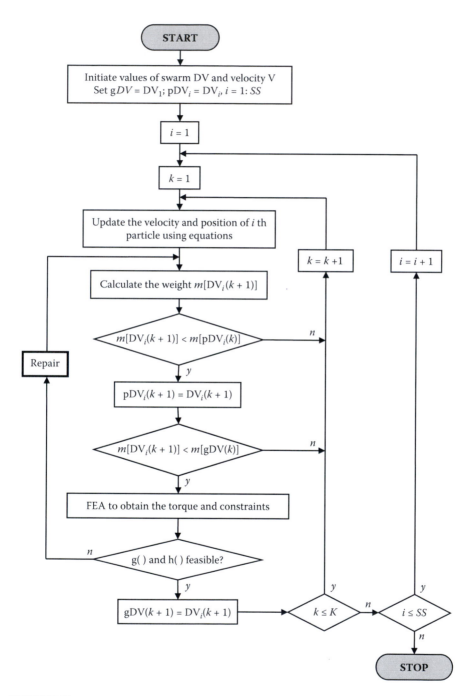

FIGURE 7.10
Flowchart for optimal design of the BMR brake using FEA.

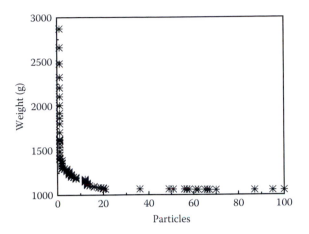

FIGURE 7.11
Convergence of the objective in the optimization problem using FEA. (From Nguyen, P.B. et al., *Smart Materials and Structures*, 20, 12, 2011. With permission.)

(a) Components of the BMR brake

(b) Assembled device

FIGURE 7.12
Photograph of a manufactured BMR brake with optimal dimensions.

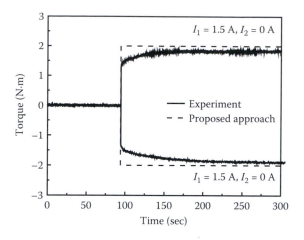

FIGURE 7.13
Torque values obtained from experiment and simulation with the proposed approach. (From Nguyen, P.B. et al., *Smart Materials and Structures*, 20, 12, 2011. With permission.)

other is turned off. The obtained torque values as shown in Figure 7.13 for two directions are stabilized approximately 1.88 Nm and 1.92 Nm, respectively. It is realized that these values are lower than the desired one in the optimal design (2 Nm). This might result from imperfections in manufacturing and assembly as well as the effect of dry friction. However, this difference is not very significant.

7.3 Torsional MR Brake

7.3.1 Control System of Torsional Vibration

Figure 7.14 shows a configuration of torsional vibration control of a rotating shaft system using the MR brake absorber. A shaft system (the shaft and its relating components) is driven by a variable torque T_S. The moment of inertia and torsional stiffness of the shaft system are, respectively, J_{so} and K_S. Because the shaft system is not perfectly rigid, it will vibrate about the axis of rotation under the variable driving torque T_S, especially at its resonant frequency. In order to reduce the vibration of the shaft system, an MR brake absorber is attached to the end of the shaft. The housing of the MR brake is fixed to the shaft rotating together with the shaft system. The total moment of inertia of the shaft system (including the ME brake housing) is denoted by J_S. The disc of the brake is placed inside the housing, and freely rotates relative to the housing, acting as a dynamic torsional vibration absorber. The gaps between the housing and the disc are filled with MR fluid. The vibration energy of the shaft system is absorbed by the disc of the brake,

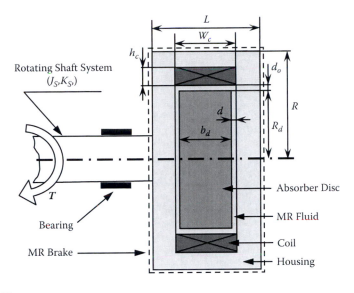

FIGURE 7.14
Configuration of an MR brake absorber for torsional vibration control.

that is, it is converted into the vibration energy of the disc through the frictional torque of the MR brake. If the friction torque is extremely small, very little vibration energy of the shaft is transferred to the disc. On the other hand, if the friction torque is very large, the disc "locks" on the housing and follows its motion. There is then no relative slip and hence no energy dissipation. Thus, there exists an optimal value of the friction torque, which maximizes the dissipated energy of the shaft vibration. In this configuration, the optimal friction torque is achieved by the current applied to the coil of the brake. In this work, a disc-type MR brake is employed and its significant geometric design variables are shown in Figure 7.14.

The key question in the design of the MR fluid brake is to establish a relationship between the braking torque and the parameters of the structure and the magnetic field strength. As is well known, the MR fluid behavior is well predicted by using the constitutive Bingham model, especially for MR fluid devices operating in low shear rates. Because the MR fluid gap size is very small, it is reasonable to assume a linear shear rate distribution of the MR fluid in the gaps. By applying the Bingham plastic model, the friction torque of the MR brake can be approximately determined by [28]

$$T_b = T_{ye} + T_{\mu e} + T_a = \frac{4\pi\tau_y}{3}R_d^3 + \frac{\pi\mu R_d^4\Omega}{d} + 2\pi R_d^2 b_d \left[\tau_{y0} + \mu_0\left(\frac{\Omega R_d}{d_o}\right)\right] \quad (7.42)$$

where μ *and* τ_y are, respectively, the post-yield viscosity and yield stress of the MR fluid in gaps at the end-faces of the disc, which depend on the applied current; μ_o and τ_{y0} are, respectively, the post-yield viscosity and yield stress of the

MR fluid when no magnetic field is applied; R_d and b_d are the radius and width of the disc; d is the width of the MR fluid ducts at the end-faces of the disc; d_o is the width of the annular duct of MR fluid; and Ω is the relative angular velocity between the housing and the disc of the MR brake. T_{ye} and $T_{\mu e}$ are friction torques at the end-faces of the disc due to yield stress and post-yield viscosity of the MR. T_a is the friction torque at the cylindrical face of the disc. In most cases, it is noteworthy that the value of the first term in Equation (7.42) is much greater than that of the second and the third term. Therefore, the MR friction torque of the MR brake is essentially similar to a dry friction force.

In case no power is applied to the coil, the friction torque of the MR brake (zero-field friction torque) can be approximately determined as

$$T_{b0} = 2\pi\tau_{y0}R_d^2\left(\frac{2R_d}{3} + b_d\right) + \Omega\pi\mu_0 R_d^3\left(\frac{R_d}{d} + \frac{2b_d}{d_o}\right) \tag{7.43}$$

In this work, the commercial MR fluid (MRF132-DG) made by Lord Corporation is used. The post-yield viscosity of the MR fluid is assumed to be independent on the applied magnetic field, $\mu = \mu_0 = 0.1\ Pa.s$. The induced yield stress of the MR fluid as a function of the applied magnetic field intensity (H) is shown in Figure 7.15. By applying the least square curve fitting method, the yield stress of the MR fluid can be approximately expressed by

$$\tau_y = C_0 + C_1 H + C_2 H^2 + C_3 H^3 \tag{7.44}$$

In Equation (7.44), the unit of the yield stress is kPa while that of the magnetic field intensity is kA/m. The coefficients C_0, C_1, C_2, and C_3 are identified,

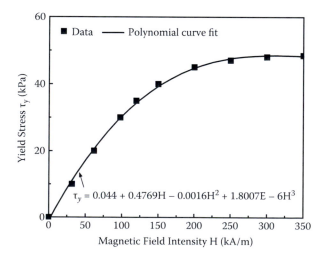

FIGURE 7.15
Yield stress of MR fluid as a function of magnetic field intensity. (From Nguyen, Q.H. et al., *Smart Materials and Structures*, 21, 2, 2012. With permission.)

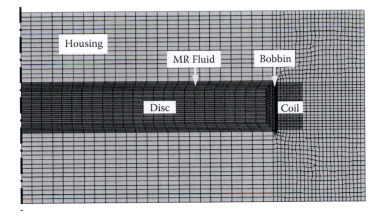

FIGURE 7.16
Finite element models to analyze the magnetic circuit of the MR brake. (From Nguyen, Q.H. et al., *Smart Materials and Structures*, 21, 2, 2012. With permission.)

respectively, as 0.044, 0.4769, −0.0016, and 1.8007E-6. It is observed from the figure that when the magnetic field intensity is up to 300 kA/m or higher, the MR fluid reaches a saturation state and the saturated yield stress is 48.5 kPa. In order to calculate the damping force of the MR damper, it is necessary to solve the magnetic circuit of the brake. In the analysis, a finite element method is applied to solve the magnetic circuit. The finite element model of the brake using the 2D-axisymmetric couple element (PLANE 13) of the commercial ANSYS software is shown in Figure 7.16. Because the geometric dimensions of the MR brake are altered during the optimization process, the meshing size is assigned by the number of segments for each line. Obviously the finer the meshing is, the more accurate the solution can be obtained, but the computation time is increased. As the meshing is fine enough, the solution is converged. The number of elements is 1766. When the number of elements is increased to 3079, the solution is changed 0.062% compared to that in the case of 1766 elements. However, the computation time is increased twice. Generally, the magnetic intensity at the MR ducts is not constant and an average value of the magnetic density should be used. In this work, this average value is calculated by a numerical integration of the magnetic intensity along the gaps.

7.3.2 Optimal Design

As previously mentioned, an optimal value of the friction torque exists that maximizes the dissipated energy of the shaft vibration. Because the friction torque due to post-yield viscosity of MR fluid is normally very small compared to that due to the yield stress, in essence the friction torque is similar to a dry friction torque. Figure 7.17 shows the free body diagram of the shaft

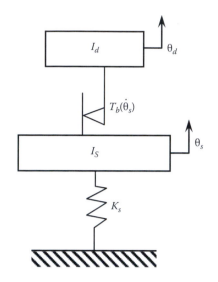

FIGURE 7.17
Free body diagram of the shaft system featuring an MR brake absorber.

system featuring the MR brake absorber. The system can be mathematically expressed as :

$$J_S\ddot{\theta}_S + T_b(\dot{\theta}_S) + K_S\theta_S = T_0 \sin \omega t \qquad (7.45)$$

$$J_d\ddot{\theta}_d = T_b(\dot{\theta}_S) \qquad (7.46)$$

In the above equation, J_S and K_S are the moment of inertia and torsional stiffness of the shaft system including the housing of the MR brake. J_d is the moment of inertia of the brake disc. θ_S and θ_d are, respectively, the angular displacement of the shaft and the disc. $T_b(\dot{\theta}_S)$ is the friction torque of the MR brake. By ignoring the minor friction torque due to the post-yield viscosity of the MR fluid, $T_b(\dot{\theta}_S)$ can be approximately calculated by

$$T_b(\dot{\theta}_S) = T_b \operatorname{sgn}(\dot{\theta}_S - \dot{\theta}_d) \cong \frac{4\pi\tau_y}{3} R_d^3 \operatorname{sgn}(\dot{\theta}_S - \dot{\theta}_d) \qquad (7.47)$$

It was shown by Den Hartog and Omondroyd [17] that the optimal value of the dry friction torque (the torque that maximizes the dissipated energy of the shaft vibration) can be determined by

$$T_{opt} = \frac{2}{\pi} J_d\omega^2\Theta \qquad (7.48)$$

where Θ is the amplitude of the shaft torsional vibration. In order to determine Θ, it is assumed that the vibration of the shaft takes the form $\theta_S = \Theta \sin(\omega t + \varphi)$, where φ is a phase lag. In this case, the work per cycle of the optimal friction torque T_{opt} can be determined by

$$W_{cyc} = \Theta \int_0^{2\pi} T_{opt} \cos(\omega t) d(\omega t) \qquad (7.49)$$

By introducing an equivalent damping coefficient C_{eq} such that the work per cycle due to this equivalent damping coefficient equals that of optimal friction torque T_{opt}, the following equation holds:

$$\Theta \int_0^{2\pi} T_{opt} \cos(\omega t) d(\omega t) = \Theta^2 C_{eq} \omega \int_0^{2\pi} \cos^2(\omega t) d(\omega t) = \Theta^2 C_{eq} \omega \pi \qquad (7.50)$$

From Equation (7.50), the equivalent damping coefficient C_{eq} is determined by

$$C_{eq} = \frac{\int_0^{2\pi} T_{opt} \cos(\omega t) d(\omega t)}{\Theta \omega \pi} = \frac{4|T_{opt}|}{\Theta \omega \pi} = \frac{4\sqrt{2}}{\pi^2} J_d \omega \qquad (7.51)$$

By assuming that the effect of the friction torque of the MR brake to the torsional vibration of the shaft is similar to that of a viscous damping torque whose damping coefficient is C_{eq}, the amplitude of the torsional vibration of the shaft can be derived by solving Equation (7.45) as [16, 17]:

$$\Theta = \frac{T_0}{K_S} \frac{1}{\sqrt{\left(1 - \dfrac{\omega^2}{\omega_n^2}\right)^2 + \left(\dfrac{C_{eq}\omega}{K_s}\right)^2}} = \frac{\Theta_{st}}{\sqrt{\left(1 - \dfrac{\omega^2}{\omega_n^2}\right)^2 + \dfrac{32}{\pi^4}\left(\dfrac{J_d\omega^2}{K_s}\right)^2}} \qquad (7.52)$$

where $\Theta_{st} = T_0/K_S$ is the static torsional displacement of the shaft system, and ω_n is the natural frequency of the shaft system, $\omega_n = \sqrt{K_S/J_S}$. By defining $\mu_j = J_d/J_s$ and $r = \omega/\omega_n$, Equation (7.52) can be rewritten as

$$\Theta = \frac{\Theta_{st}}{\sqrt{(1 - r^2)^2 + \dfrac{32\mu_j^2 r^4}{\pi^4}}} \qquad (7.53)$$

Equation (7.53) expresses the amplitude of the shaft system vibration as a function of the frequency ratio r. The maximum value of Θ can determined by the condition $d\Theta/dr = 0$, which is given in the following results:

$$\Theta_{max} = \Theta_{st} \sqrt{1 + \frac{\pi^4}{32\mu_j^2}}, \text{ at } r = \frac{\pi^2}{\sqrt{\pi^4 + 32\mu_j^2}} \qquad (7.54)$$

It is noteworthy from Equation (7.54) that the maximum dynamic amplification of the shaft vibration Θ/Θ_{st} depends only on the ratio of the moment of inertia. Therefore, the maximum vibration of the shaft can be controlled by applying an appropriate absorber. In the design of the vibration absorber system, normally the value of the ratio of the moment of inertia μ_J is chosen from 0.25 to 0.5. The corresponding maximum dynamic amplification ranges from 7.05 to 3.63. The smaller value of μ_J causes an acute increase of dynamic amplification while the higher value of μ_J results in a large and heavy structure of the brake.

Plug Θ from Equation (7.53) into Equation (7.48). Then, the optimal friction torque can be expressed as

$$T_{opt} = \frac{2\mu_J T_0}{\pi} \frac{r^2}{\sqrt{(1-r^2)^2 + \frac{32\mu_J^2 r^4}{\pi^4}}} \tag{7.55}$$

From Equation (7.55), it is straightforward to show that the maximum value of the optimal friction torque is determined by

$$T_{opt,max} = \frac{\sqrt{2}\pi T_0}{4}, \text{ at } r = 1 \tag{7.56}$$

It is noted that the maximum optimal friction torque determined by Equation (7.56) is also the optimal friction torque mentioned by Den Hartog [17, 18] and Ye [19]. In order to validate the optimal friction torque expressed by Equation (7.55), the simulation of the shaft system integrated with a controllable dry friction absorber is performed and the results are presented. Figure 7.18 shows the swept sinusoidal response of the shaft system employing a controllable dry friction absorber. The excitation torque in this case is expressed by

$$T_S = T_0 \sin\left(K_T e^{(-t/L_T)} - 1\right) \tag{7.57}$$

where

$$K_T = \frac{\omega_1 t_\Sigma}{\ln(\omega_2/\omega_1)};$$

$$L_T = \frac{t_\Sigma}{\ln(\omega_2/\omega_1)};$$

t_Σ is the length of the excitation,

$$t_\Sigma = \frac{2^N}{f_s};$$

$\omega_1(s^{-1})$ and $\omega_2(s^{-1})$ are the starting and ending frequency to be swept;
f_s is the sampling rate (Hz); and
N is the an integral number disseminating the length of the excitation.

The higher value of N causes an increase in the length of the excitation t_Σ, which results in a large computation time and a huge number of data. However, smaller values of N may cause a bad result. In this work, N is set by 13 based on a trial-and-error method. The swept sinusoidal excitation torque with the amplitude $T_0 = 5$ Nm is pictorially shown in Figure 7.18(a). The maximum value of the optimal friction torque calculated from Equation (7.56) is 5.553 Nm. For this simulation, a shaft system, whose moment of inertia is $J_{so} = 50$ kgcm² and torsional stiffness is $K_S = 26000$ N, is considered. Of note, the housing of the absorber is fixed to the shaft, which increases the moment of inertia of the shaft system. Thus, the moment of inertia of the shaft system is $J_S = J_{so} + J_h$, where J_h is the moment of inertia of the absorber housing. The ratio of moment of inertia is set by $\mu_J = 0.25$ and 0.5. In the simulation, the following six cases are considered:

- Case 1: $J_h = 20\%$ of J_{so} and optimal torque, $T_b(\dot{\theta}_S) = T_{opt}$
- Case 2: $J_h = 20\%$ of J_{so} and under optimal torque, $T_b(\dot{\theta}_S) = 0.88T_{opt}$
- Case 3: $J_h = 20\%$ of J_{so} and over optimal torque, $T_b(\dot{\theta}_S) = 1.25T_{opt}$
- Case 4: $J_h = 20\%$ of J_{so} and constant optimal torque, $T_b(\dot{\theta}_S) = T_{opt,max}$

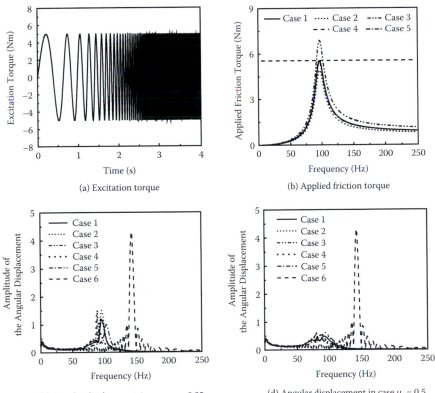

(a) Excitation torque

(b) Applied friction torque

(c) Angular displacement in case $\mu_J = 0.25$

(d) Angular displacement in case $\mu_J = 0.5$

FIGURE 7.18

Swept sinusoidal responses of the shaft system employing a controllable dry friction absorber. (From Nguyen, Q.H. et al., *Smart Materials and Structures*, 21, 2, 2012. With permission.)

- Case 5: $J_h = 40\%$ of J_{so} and optimal torque, $T_b(\dot{\theta}_S) = T_{opt}$
- Case 6: without the absorber

The friction torque for the abovementioned cases are presented in Figure 7.18(b). It is observed from Figure 7.18(c) and Figure 7.18(d) that the torsional vibration of the shaft is significantly suppressed by the absorber. Among the cases considered, the first case [optimal torque determined by Equation (7.55)] provides the best control of the shaft vibration. More specifically, a comparison between the first case (frequency-dependent optimal torque) and the fourth case (constant optimal torque mentioned by Den Hartog [17, 18] and Ye [19]) shows that the resonant frequency of the system in the first case is higher than that in the fourth case while the magnitude of the resonant peak of the former is significantly smaller (around 70%) than that of the latter. It is also observed that the vibration of the shaft system is affected by the moment of inertia of the absorber housing (J_h). The smaller the value of J_h, the better the vibration control of the shaft. In addition, a comparison of Figure 7.18(c) and Figure 7.18(d) shows that the larger the value of the moment of inertia μ_J is, the smaller the vibration of the shaft can be obtained. However, a large value of μ_J generally results in a heavy weight, large space, and high cost absorber. As previously mentioned, because the housing of the MR absorber is fixed to the shaft, this can cause the vibration of the shaft to be more severe. Therefore, in the optimal design of the MR damper, the moment of inertia of the housing should be minimized. In addition, the smaller the moment of inertia of the housing is, the smaller the size and the lower the cost of the MR brake, in general.

Taking all of the above into consideration, the optimal design of the MR brake for the torsional vibration absorber can be summarized as: Find the optimal value of significant dimensions of the MR brake such as the coil width W_c, the coil height h_c, the gap size d and d_o, the disc radius R_d, the disc width b_d, the thickness of the side housing t_h, and the thickness of the cylindrical housing t_o that minimized the moment of inertia of the brake housing, subjected to the following constraints:

I. The maximum braking torque is greater than the maximum value of the optimal friction torque.

$$T_b \cong \frac{4\pi\tau_y}{3} R_d^3 \geq \frac{\pi\sqrt{2}T_0}{4} \tag{7.58}$$

II. The ratio of the moment of inertia is greater than a specific design value.

$$\mu_J = \frac{J_d}{J_{so} + J_h} \geq \mu_d \tag{7.59}$$

where μ_d is the design ratio of the moment of inertia, the ratio J_{so} is the moment of inertia of the shaft without the brake, and J_h is the moment of inertia of the brake housing.

In order to obtain the optimal solution, a FEA code integrated with an optimization tool is employed. The detailed procedure to obtain the optimal solution of magnetorheological fluid (MRF) devices based on FEA has been mentioned in several publications [29,30].

7.3.3 Results and Discussions

Torsional vibrations of a rotating shaft system are considered in this section. An optimal design of the MR brake for the torsional absorber is performed based on the optimization problem. It is assumed that the brake disc and housing are made of commercial silicon steel, the coil wires are sized as 21-gauge (diameter = 0.511 mm) whose allowable working current is 2.5 A. A shaft system, whose moment of inertia is $J_{so} = 50$ kgcm² and torsional stiffness is $K_S = 26000$ Nm, is considered. The shaft is assumed to vibrate due to a harmonic excitation with the amplitude $T_0 = 5$ Nm. The maximum value of the optimal friction torque calculated from Equation (7.56) is 5.553 Nm.

Figure 7.19 shows the optimal solution of the MR brake when the ratio of the moment of inertia is set by $\mu_d = 0.25$. It is noted that the smaller the value of d and d_o, the higher the value of the friction torque can be achieved. However, this causes a higher value of the zero-field friction torque of the MR brake, which may reduce performance of the MR brake absorber, especially at frequencies far away from the resonance. Furthermore, the manufacturing cost is also an important issue to be considered. In the analysis, the size of the MR ducts d and d_o are chosen as 1 mm. For manufacturing advantages, the coil width is determined by $W_c = b_d + 2d$. The other significant dimensions such as the coil height h_c, the disc radius R_d, the disc thickness b_d, the thickness of the side housing t_h, and the thickness of the cylindrical housing t_o are considered as design variables. From the figure, it can be found that with a convergence tolerance of 0.5%, the optimal process is converged after 31 iterations and solution at the 31st iteration is considered the optimal one. The optimal values of h_c, R_d, b_d, t_h, and t_o are, respectively, 1.1 mm, 52.6 mm, 18.6 mm, 3.35 mm, and 3.25 mm. It is noted that the optimal value of h_c is equal to its lower limit in this case. It is also observed from the figure that the braking torque of the optimized brake and the ratio of moment of inertia are greater than the constrained values, which are 5.553 Nm and 0.25, respectively. The moment of inertia of the MR brake housing at the optimum is 19 kgcm², which is around 38% of the shaft system. From the results, it can be found that the outer radius R and the length L of the optimized MR brake are, respectively, 58.2 mm and 27.3 mm. The braking torque and power consumption as functions of the applied current of the optimized MR brake are shown in Figure 7.19(c). From the figure, it is observed that the braking

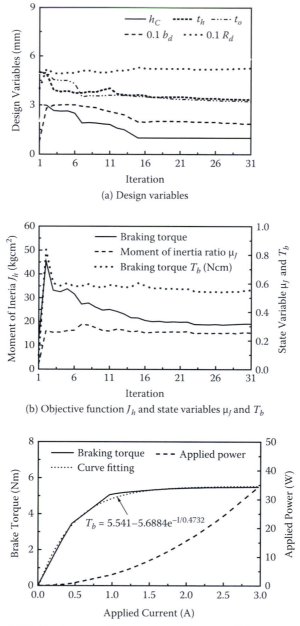

(a) Design variables

(b) Objective function J_h and state variables μ_J and T_b

$$T_b = 5.541 - 5.6884e^{-I/0.4732}$$

(c) Braking torque and power consumption vs. applied current

FIGURE 7.19
Optimization solution of the MR brake absorber in case $\mu d = 0.25$. (From Nguyen, Q.H. et al., *Smart Materials and Structures*, 21, 2, 2012. With permission.)

torque almost reaches saturation when the applied current is higher than 1.6 A. Therefore, in order to reduce the power consumption and the potential heat of the MR brake, the applied current should be limited by 1.6 A. By applying a least square curve-fitting algorithm, the braking torque of the MR brake can be expressed as:

$$T_b = 5.541 - 5.6884e^{-\frac{I}{0.4732}} \tag{7.60}$$

From Equation (7.60), the applied current I to the MR brake coil in order to produce a braking torque T_b can be inversely determined by

$$I = -0.4732 \log\left(\frac{5.541 - T_b}{5.6884}\right) \tag{7.61}$$

Figure 7.20 shows the optimal solution of the MR brake when the ratio of the moment of inertia is set by $\mu_d = 0.5$. From the figure, it can be found that with a convergence tolerance of 0.5%, the optimal process is converged after 26 iterations and the solution at the 26th iteration is considered as the optimal one. The optimal values of h_c, R_d, b_d, t_h, and t_o are, respectively, 1.1 mm, 55.5 mm, 33.5 mm, 3.1 mm, and 3 mm. The optimal value of h_c is also equal to its lower limit in this case. The braking torque of the optimized brake and the ratio of moment of inertia are greater than the constrained values, which are 5.553 Nm and 0.5, respectively. The moment of inertia of the MR brake housing at the optimum is 27.7 kgcm², which is around 55.4% of the shaft system. In this case, the outer radius R and the length L of the optimized MR brake are, respectively, 61 mm and 41.6 mm. From Figure 7.20(c), it is observed that the braking torque almost reaches to saturation when the applied current is higher than 1.5 A. Therefore, the applied current should be limited by 1.5 A. By applying a least square curve-fitting algorithm, the braking torque of the MR brake can be expressed as:

$$T_b = 5.488 - 5.6624e^{-\frac{I}{0.2604}} \tag{7.62}$$

From Equation (7.62), the applied current $I(A)$ to the MR brake coil in order to produce a braking torque $T_b(Nm)$ can be inversely determined by

$$I = -0.2604 \log\left(\frac{5.488 - T_b}{5.6624}\right) \tag{7.63}$$

In order to evaluate the optimized MR brake, simulated results are obtained for the rotating shaft system employing the optimized MR brake absorbers. The following five cases are considered: (1) Optimal: the current applied to the coil is determined such that the braking torque generated by the MR brake is equal to the optimal friction torque, $T_b = T_{opt}$ determined by Equation (7.55). (2) Under Opt.: the current is applied to the coil to produce a

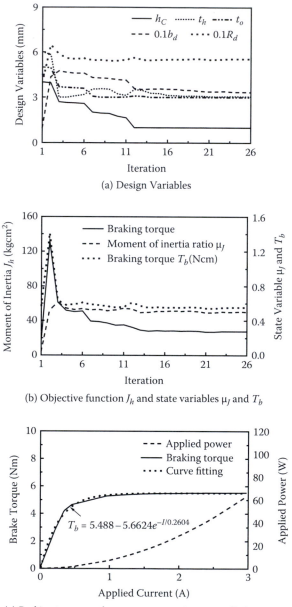

(a) Design Variables

(b) Objective function J_h and state variables μ_J and T_b

$$T_b = 5.488 - 5.6624e^{-I/0.2604}$$

(c) Braking torque and power consumption vs. applied current

FIGURE 7.20
Optimization solution of the MR brake absorber in case $\mu d = 0.5$. (From Nguyen, Q.H. et al., *Smart Materials and Structures*, 21, 2, 2012. With permission.)

braking torque smaller than the optimal friction torque, $T_b = 0.88T_{opt}$. (3) Over Opt.: the current is applied to the coil to produce a braking torque higher than the optimal friction torque, $T_b = 1.25T_{opt}$. (4) Const. Opt.: a constant current is applied to the coil to generate a braking torque equal to the maximum optimal friction torque, $T_b = T_{opt,max}$. (5) No Absorb.: no absorber is employed.

Figure 7.21 shows the swept sinusoidal response of the shaft system employing the optimized MR brake in the case of $\mu_d = 0.5$. The results show that the torsional vibration of the shaft is significantly suppressed by the

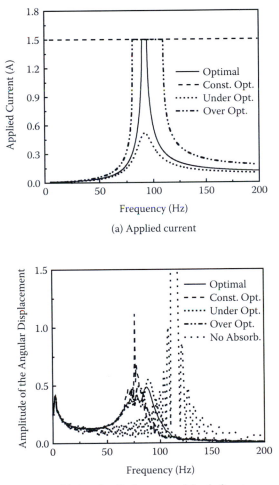

(a) Applied current

(b) Angular displacement of the shaft system

FIGURE 7.21

Swept sinusoidal responses of the shaft system employing the optimized MR brake absorber ($\mu d = 0.5$) in the frequency domain. (From Nguyen, Q.H. et al., *Smart Materials and Structures*, 21, 2, 2012. With permission.)

optimized MR absorber. Among the cases considered, the first case [braking torque determined by Equation (7.55)] provides the best control of the shaft vibration. Again, a comparison between the first case (frequency-dependent optimal torque) and the fourth case (constant optimal torque mentioned by Den Hartog [17, 18] and Ye [19]) shows that the resonant frequency of the system in the first case is higher than that in the fourth case while the magnitude of the resonant peak of the former is significantly smaller (around 45%) than that of the latter. Figure 7.22 shows the swept sinusoidal response of the shaft system featuring the optimized MR brake in the case of $\mu_d = 0.25$. The results also show that the torsional vibration of the shaft is significantly suppressed

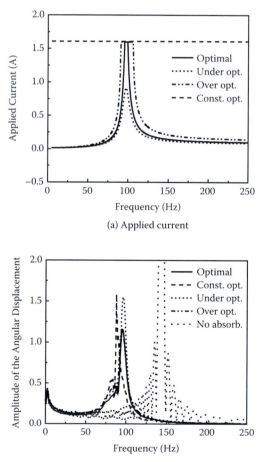

(a) Applied current

(b) Angular displacement of the shaft system

FIGURE 7.22

Swept sinusoidal responses of the shaft system employing the optimized MR brake absorber ($\mu d = 0.25$) in the frequency domain. (From Nguyen, Q.H. et al., *Smart Materials and Structures*, 21, 2, 2012. With permission.)

by the optimized MR absorber and the first case provides the best vibration control performance. A comparison of Figure 7.21 and Figure 7.22 shows that the vibration control performance of the optimized MR brake absorber in the case of $\mu_d = 0.5$ is better than that of the optimized MR brake absorber in the case of $\mu_d = 0.25$ in the vicinity of the system resonance. However, at the frequencies far away from the resonance, the latter is somewhat better. The reason is that the moment of inertia of the housing in the case of $\mu_d = 0.5$ is higher than that in the case of $\mu_d = 0.25$. This large moment of inertia causes the vibration of the shaft to become more severe. Furthermore, the high cost, the available space, and the heavy weight of the MR brake are also critical issues that should be taken into account when designing the MR brake absorber.

7.4 Some Final Thoughts

This chapter highlighted MR brake systems for braking function or vibration-absorbing functions of vehicles. Section 7.2 first described a BMR brake with an engineering optimization problem. After the introduction and mechanical modeling of the BMR brake, its magnetic circuit was discretized to numerous elements from which the model for this circuit was proposed. To assess the accuracy of this magnetic circuit method, an optimal design with the integrated proposed method was undertaken. In the work, a PSO algorithm, which is known to be effective in solving problems with numerous constraints and multi-local optima, in combination with a gradient-based repair method was proposed. The optimal solution was then analyzed and verified using Maxwell software. It was shown that the maximum percentage error between the optimal solution and the Maxwell solution is 4% for the braking torque. Moreover, an alternative optimization using FEA was also conducted for comparing with the proposed optimal design. It was shown that the results in both optimization problems were insignificantly different. Moreover, it was realized that the computation time of the optimal design using the proposed magnetic circuit analysis was drastically faster than that using FEA. The experimental validation with the manufactured device based on the optimal dimensions showed that there was not much significant difference between the value obtained via experiment and the proposed one.

Section 7.3 discussed an MR brake absorber for torsional vibration control of a rotating shaft. A configuration of torsional vibration control of a rotating shaft with an MR brake absorber was first proposed, and braking torques of the MR brake were derived based on the Bingham plastic model of the MR fluid. By assuming that the nature of the braking torque of the MR brake is similar to that of a dry friction torque, the optimal value of the braking torque has been derived and the optimal design problem of the MR brake has been developed. The optimal design of the MR brake absorber for a specific rotating shaft system was obtained based on the ANSYS finite element analysis of the MR brake magnetic circuit and first order optimization method with the golden-section algorithm. The simulation results of the system excited by a swept sinusoidal torque showed that the torsional vibration of the shaft can be significantly suppressed by the optimized MR absorber and the optimal braking torque provides the best control performance of the shaft vibration. The results also have shown that by choosing a higher value of moment of inertia ratio, the vibration control performance of the rotating system is better in the vicinity of the system resonant domain. However, at the frequencies far away from the resonance, the vibration of the system is more severe.

References

[1] Webb, G. M. 1998. Exercise apparatus and associated method including rheological fluid brake. U.S. Patent, No. 5,810,696.

[2] Avraam, M., Horodinca, M., Romanescu, I., and Preumont, A. 2010. Computer controlled rotational MR-brake for wrist rehabilitation device. *Journal of Intelligent Material Systems and Structures* 21: 1543–1557.

[3] Blake, J. and Gurocak, H. B. 2009. Haptic glove with MR brakes for virtual reality. *Mechatronics, Transactions on IEEE/ASME* 14: 606–615.

[4] Han, Y. M., Noh, K. W., Lee, Y. S., and Choi, S. B. 2010. A magnetorheological haptic cue accelerator for manual transmission vehicles. *Smart Materials and Structures* 19: 1–10.

[5] Li, W. H., Liu, B., Kosasih, P. B., and Zhang, X. Z. 2006. Development of an MR-brake-based haptic device. *Smart Materials and Structures* 15: 1960–1966.

[6] Park, E. J., Luz, L. F., and Suleman, A. 2008. Multidisciplinary design optimization of an automotive magnetorheological brake design. *Computers & Structures* 86: 207–216.

[7] Karakoc, K., Park, E. J., and Suleman, A. 2008. Design considerations for an automotive magnetorheological brake. *Mechatronics* 18: 434–447.

[8] Li, W. H. and Du, H. 2003. Design and experimental evaluation of a magnetorheological brake. *International Journal of Advanced Manufacturing Technology* 21: 508–515.

[9] Gudmundsson, K. H., Jonsdottir, F., and Thorsteinsson, F. 2010. A geometrical optimization of a magneto-rheological rotary brake in a prosthetic knee. *Smart Materials and Structures* 19: 1–11.

[10] Nguyen, Q. H. and Choi, S. B. 2010. Optimal design of an automotive magnetorheoloogical brake considering geometric dimensions and zero-field friction heat. *Smart Material and Structures* 19: 1–11.

[11] Nguyen, P. B. and Choi, S. B. 2011. A new approach to magnetic circuit analysis and its application to optimal design of a bi-directional magnetorheological brake. *Smart Materials and Structures* 20: 1–12.

[12] Nguyen, Q. H. and Choi, S. B. Optimal design of magnetorheological brake absorber for torsional vibration control. *Smart Materials & Structures* 21: 1–9.

[13] Taraza, D. and Wölfel, H. 1995. Torsional vibration control by means of an active vibration absorber. *SAE Technical Papers*: 952806.

[14] Ostman, F. and Toivonen, H. T. 2008. Model-based torsional vibration control of internal combustion engines. *IET Control Theory & Applications* 11: 1024–1032.

[15] Wei, Y. D., Lu, Y. G., and Chen, Z. 2005. Active control of torsional vibration of flexible bars. *International Journal of Information and Systems Sciences* 1: 347–354.

[16] Silva, C. W. 2000. *Vibration: Fundamentals and Practices*. Boca Raton, FL: CRC Press.

[17] Den Hartog, J. P. 1956. *Mechanical Vibrations*, 4th ed., New York: McGraw-Hill.

[18] Den Hartog, J. P. and Ormondroyd, J. 1929. Torsional vibration dampers, *Transactions of American Society of Mechanical Engineers APM* 52: 133–52.

[19] Ye, S. and Williams, K. A. 2006. Torsional friction damper optimization. *Journal of Sound and Vibration* 294: 529–546.

[20] Ye, S. and Williams, K. A. 2004. Semi-active control of torsional vibrations using a MR fluid brake. *Proceedings of SPIE* 5390: 135–146.

[21] Ye, S. and Williams, K. A. 2005. Torsional vibration control with a MR fluid brake. *Proceedings of SPIE* 5760: 283–292.

[22] Sun, Y. and Thomas, M. 2010. Control of torsional rotor vibrations using an electrorheological fluid dynamic absorber. *Journal of Vibration and Control* 17: 1253–1264.

[23] Eberhart, R. and Kennedy, J. 1995. Particle swarm optimization. *Proceedings of IEEE International Conference on Evolutionary Computation* 1942–1948.

[24] Shi, Y. and Eberhart, R. C. 1998. A modified particle swarm optimizer. *Proceedings of IEEE International Conference on Evolutionary Computation.* 69–73

[25] Chootinan, P. and Chen, A. 2006. Constraint handling in genetic algorithms using a gradient-based repair method. *Computers and Operations Research* 33: 2263–2281.

[26] http://www.lord.com

[27] http://www.ansoft.com

[28] Nguyen, Q. H. and Choi, S. B. 2010. Optimal design of an automotive magnetorheological brake considering geometric dimensions and zero-field friction heat. *Smart Materials and Structures* 19: 1–11.

[29] Nguyen, Q. H., Han, Y. M., Choi, S. B., and Wereley, N. M. 2007. Geometry optimization of MR valves constrained in a specific volume using the finite element method. *Smart Materials and Structures* 16: 2242–2252.

[30] Nguyen, Q. H. and Choi, S. B. 2008. Optimal design of vehicle MR damper considering damping force and dynamic range. *Smart Materials and Structures.* 18: 1–10.

8

Magnetorheological (MR) Applications for Heavy Vehicles

8.1 Introduction

Numerous heavy industrial vehicle types, extended driving hours, and exposure to severe working environments are transportation-related factors in industry and agriculture. The increased number of these vehicles inevitably results in increased demand for improved ride comfort. The ride vibration detected at the driver's seat significantly influences driver fatigue and safety. In fact, the ride vibration levels of commercial vehicles are 9 to 16 times higher than those of passenger cars, and most commercial vehicle drivers are exposed to this ride vibration for 10 to 20 hours a day [1]. Although ride comfort improvements have been attempted through appropriate tires and primary and secondary (cabin) suspensions, commercial vehicle drivers still suffer from the effects of low-frequency and high-amplitude ride vibration.

Furthermore, some advanced heavy vehicles require a high-performance engine to generate high power levels, which creates a tremendous amount of heat. Thus, a high-performance cooling system is one of the most important parts of advanced vehicles. When the engine is cool or even at normal operating temperature, the fan clutch partially disengages the mechanically driven radiator cooling fan, which is generally located at the front of the water pump and is driven by a belt and pulley connected to the engine crankshaft. This mechanism saves power because the engine does not have to fully drive the fan. However, conventional fan clutches have difficulty adapting to high-performance vehicles. For example, if the engine temperature rises above the desired temperature set for clutch engagement, the fan becomes fully engaged, thus drawing a higher volume of ambient air through the vehicle radiator, which in turn maintains or lowers the engine coolant temperature to an acceptable level. Most commercial fan clutches are viscous or "fluid" couplings combined with a bi-metallic sensory system similar to that in a thermostat. These fan clutches are normally quite reliable, but sometimes they fail. A common symptom of fan clutch failure is overheating in idle or heavy traffic. A poor fan clutch can also lower air-conditioning performance because the fan also cools the air-conditioner condenser, which is located in front of the radiator. Another potential symptom of fan clutch failure (always drawing air at a high rate) in a cold weather climate is that the heating system blows lukewarm air and never delivers sufficient hot air.

In this chapter, we focus on two MR applications that are available to commercial heavy vehicles: the fan clutch for automotive engine cooling systems [2], and the damper for seat suspension systems, which is normally employed for commercial vehicles as a simple and effective way to attenuate unwanted vibrations in the driver's seat.

Section 8.2 introduces a controllable fan clutch, a thermostatic device that is an integral component of automotive cooling systems. One attractive candidate for achieving this goal is to use a controllable fan clutch featuring MR fluid. Neelakantan and Washington [3] proposed the use of MR transmission clutches for automotive applications to reduce centrifuging. Kavlicoglu et al. [4] suggested an MR fluid-limited slip differential clutch to transmit power to the driving wheels through a differential system. Hakogi et al. [5] used MR fluid clutches to control torque for rehabilitation teaching robots. Choi et al. [6] reported a performance comparison of field-controlled characteristics between electrorheological (ER) and MR clutches. Single-disc ER and MR clutches were used for the study and closed-loop feedback controllers were used for torque regulation and control tracking. Benetti and Dragoni [7] performed a nonlinear magnetic analysis of multi-plate MR brakes and clutches.

This section experimentally demonstrates the effectiveness of an MR fan clutch for controlling a vehicle engine's cooling system temperature. To achieve this goal, an appropriate size (practically usable) of the controllable fan clutch using MR fluid is devised and the control performance of the vehicle engine cooling system is experimentally evaluated by activating an MR fan clutch. The drum-type MR clutch is designed by considering design parameters of conventional fan clutches, and its characteristic model is derived based on a Bingham plastic model of MR fluid. Subsequently, optimization to determine principal design parameters such as housing width is performed using a finite element method (FEM). A robust sliding mode controller is then formulated by treating the time constant of the fan clutch system as an uncertain parameter. The controller associated with the MR fan clutch is experimentally realized for the cooling system, and active feedback control performances such as temperature controllability are evaluated and compared with the passive control results.

Section 8.2 presents a very attractive and effective semi-active seat suspension system featuring MR fluids [8]. This type of semi-active seat suspension system has several advantages such as fast response time, continuous controllable damping force, and low energy consumption compared to active systems. Semi-active dampers and their damping force models have been investigated and applied in various suspension systems with higher yield stress characteristics [9–15]. As a first step, a cylindrical MR seat damper is devised and its damping force is experimentally evaluated with respect to input current (magnetic field) intensity. The seat damper is then integrated with the primary passive suspension to formulate a full-vehicle model, and its governing equation of motion is derived. A semi-active skyhook control algorithm for the MR seat suspension is designed to attenuate unwanted vibration in the driver's seat, and the controller is implemented via hardware-in-the-loop simulation (HILS). A commercial vehicle is equipped with a passive primary suspension, the devised MR seat suspension is tested under both bump and random road conditions, and vibration control results are presented in time and frequency domains.

8.2 MR Fan Clutch

8.2.1 Design Optimization

Generally, the yield stress of MR fluid has exponentially proportional relation to the magnetic field. This relationship is known as the Bingham plastic model and can be governed by the equation [16]:

$$\tau = \tau_y(H) + \eta\dot{\gamma}$$
$$\tau_y = \alpha H^\beta$$

$$(8.1)$$

where τ is the shear stress of MR fluid, η is the plastic viscosity (i.e., viscosity at $H = 0$), and $\dot{\gamma}$ is the fluid shear rate. H is the field intensity, which is proportional to current. $\tau_y(H)$ is the field-dependent yield stress of the MR fluid. Figure 8.1 shows the Bingham characteristics of MR fluid. The field-dependent yield stress indicates the shear stress that appears when an external magnetic field is applied. It is difficult to measure the exact value of the yield stress of MR fluid because the flow of MR fluid has complex behavior. Thus, the dynamic yield stress is defined when the shear rate is zero. Generally, the dynamic yield stress is treated as the yield stress of MR fluid and expressed by exponential function according to the external magnetic field. The proportional factor α and exponent β, indicated in Equation (8.1), are normally determined though experimental investigation. These factors are influenced by various factors such as composition rate of solvents, particles, and production environment. The MRF-132DG fluid that is produced by Lord Corporation is employed for the MR fan clutch. MRF-132DG fluid is a hydrocarbon-based MR fluid formulated for general use in controllable energy-dissipating applications. The weight ratio of MRF-132DG is 80.98%, the plastic viscosity is 0.092 ± 0.015Pa·s, and the density is 2.98 to 3.18 g/cm³. Figure 8.2 shows the measured Bingham characteristics of MRF-132DG and from the results of curve fitting,

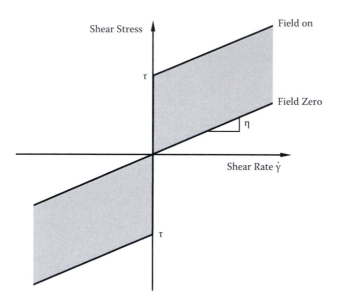

FIGURE 8.1
Bingham plastic model.

α and β are calculated. The experimentally determined Bingham model is described as:

$$\tau_y(H) = 0.1367H^{1.213} \qquad (8.2)$$

where the unit of $\tau_y(H)$ is kPa and the unit of H is A/mm.

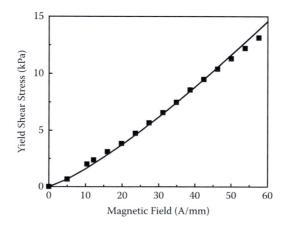

FIGURE 8.2
Measured yield stress of MR fluid. (From Kim, E.S. et al., *Smart Materials and Structures*, 19, 10, 2010. With permission.)

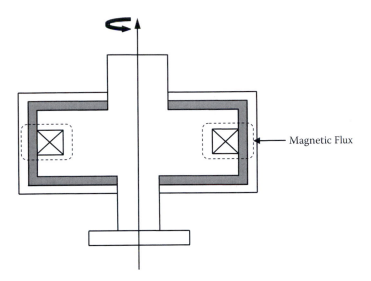

FIGURE 8.3
Motion of the MR fan clutch.

On the other hand, using the magnetic field intensity (H), the number of coil (N), and current (I), the following equation can be formulated:

$$NI = H \times L \tag{8.3}$$

where L is the length of the magnetic field path. From the motion of the MR fan clutch (refer to Figure 8.3), it can be known that the path of the magnetic field of MR fluid is two, and the path that flows through the ferromagnetic substance is one. Thus, Equation (8.3) can be expressed by

$$NI = 2 \times H_M L_M + H_S L_S \tag{8.4}$$

where the subscripts M and S indicate the MR fluid and steel, respectively. Normally, the magnetic field that flows through steel is negligible and hence Equation (8.4) can be rewritten as:

$$H_M = \frac{NI}{2g_v} = \frac{NI}{2L_M} \tag{8.5}$$

where g_v is the length of the magnetic field of MR fluid.

A drum type mechanism for the clutch is devised to transmit the shaft power to the cooling fan. Figure 8.4 shows a schematic configuration of the MR fan clutch. This roughly consists of input shaft, housing, cylindrical electromagnetic disk that has a flux guide, electromagnetic coil, and output shaft. It is assembled into the housing with a specific gap fully filled with MR fluid. A geometrical configuration of the MR fan clutch is shown in Figure 8.5. The MR fan clutch is designed to achieve rotary motion that is based on the shear

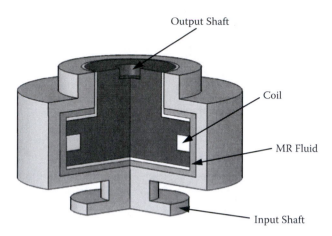

FIGURE 8.4
Schematic configuration of the proposed MR fan clutch.

mode of MR fluid. The torque is generated from rotary motion between the housing and the electromagnetic disk. The equation of the generated torque of the MR fan clutch is mathematically given by

$$T = \tau_y \times A \times R + T_\eta$$

$$= 4\pi R_o^2 t\tau_y + \frac{\pi^2 \eta \{(R_o + g_v)^4 - R_i^4\} f}{g_h} + \frac{\pi^2 \eta (R_o + g_v)^4 f}{g_h} + \frac{4\pi^2 R_o^3 \eta hf}{g_v} \quad (8.6)$$

where R_o and R_i are the outer and inner radius of the electromagnetic disk, and A is the contact area between the electromagnetic disk and MR fluid. h is

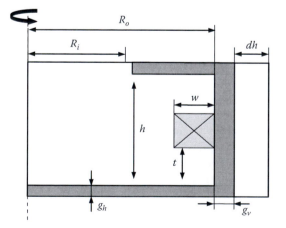

FIGURE 8.5
Geometric parameters of the MR fan clutch.

the height of the electromagnetic disk, t is the length of the electromagnetic disk except for the width of the magnetic coil, and f is the difference of angular velocity of output shaft and input shaft. g_h is the distance between the bottom of the housing and the electromagnetic disk. In addition, the friction torque from mechanical friction between parts of the MR fan clutch exists. The friction model can be written as:

$$T_f = T_{cf}\, \mathrm{sgn}(\dot{\theta})\tag{8.7}$$

where T_f is the friction torque, T_{cf} is the Coulomb friction, and $\dot{\theta}$ is the angular velocity of the MR fan clutch. Under consideration of Equation (8.7), Equation (8.6) can be expressed by:

$$T = 4\pi R_o^2 t\tau_y + \frac{\pi^2\eta\{(R_o + g_v)^4 - R_i^4\}f}{g_h} + \frac{\pi^2\eta(R_o + g_v)^4 f}{g_h} + \frac{4\pi^2 R_o^3\eta h f}{g_v} + T_{cf}\,\mathrm{sgn}(\dot{\theta})$$

$$= T_c + T_\eta + T_f\tag{8.8}$$

Thus, from Equation (8.5) and Equation (8.8), the generated torque from the MR fan clutch can be obtained by

$$T = 4\pi R_o^2 t\alpha \left(\frac{NI}{2g}\right)^\beta + \frac{\pi^2\eta\{(R_o + g_v)^4 - R_i^4\}f}{g_h}$$

$$+ \frac{\pi^2\eta(R_o + g_v)^4 f}{g_h} + \frac{4\pi^2 R_o^3\eta h f}{g_v} + T_{cf}\,\mathrm{sgn}(\dot{\theta})\tag{8.9}$$

The magnetic circuit of the MR fan clutch is optimized by the finite element analysis, for which a commercial analysis tool named ANSYS parametric design language is used. Based on a datasheet from the experimental result shown in Figure 8.2, the yield stress of the MR fluid is determined and expressed as a polynomial form:

$$\tau_y = 39.7215 \times B^4 - 132.3825 \times B^3 + 119.0925 \times B^2 + 10.281 \times B + 0.10815\tag{8.10}$$

where B is the magnetic density with a unit of T. Low carbon steel (S45C) is used as the ferromagnetic substance. Copper is used as the magnetic coil. When the optimization is performed, the reciprocal of T_c (controllable torque) is set as an objective function to maximize controllable torque. Thus, the objective function of the MR fan clutch can be expressed by

$$Obj = \min\left\{\frac{1}{T_c}\right\} = \min\left\{\frac{1}{4\pi R_o^2 t\tau_y}\right\}\tag{8.11}$$

When the magnetic circuit of the MR fan clutch is manufactured, aluminum (Al) is attached on the upper and lower sides to concentrate the magnetic flux to g_v direction. Thus, it is assumed that the magnetic flux is not dispersed in the direction of the upper and lower bounds when the optimization is performed using finite element analysis (FEA). The chosen design parameters to be optimized are the effective length of the disk except the magnetic coil (t), the depth of magnetic coil (w), and the width of housing (dh) because these geometric parameters can significantly affect the performance of the MR fan clutch. The constraints of the optimization are set as:

$$0.005 \le dh(m) \le 0.01$$

$$0.005 \le t(m) \le 0.014$$

(8.12)

Design variables are bounded as shown Equation (8.12), and the diameter and thickness of the MR fan clutch are determined to make it similar to those of commercial (conventional) fan clutches. Therefore, each value of the diameter and thickness is fixed at 140 mm and 40 mm, respectively.

The principal design parameters are optimally determined using the ANSYS parametric design language to achieve the objective function, which is to maximize the controllable torque. Because the MR fan clutch is axisymmetric, a 2D-axisymmetric coupled element is used for the electromagnetic analysis. In addition, 4-node quadrilateral meshing is used for the finite element model. The mesh size is specified by the number of elements per line rather than element size because the geometric dimensions of the MR fan clutch may vary during the optimization process. For the optimization procedure, a log file for solving the magnetic circuit of the MR fan clutch and the controllable torque is built using ANSYS parametric design language. In this log file, each design parameter is defined with its limits, tolerance, and initial value. During the optimization procedure, the geometric size can be changed, and thus the size of elements is determined according to the number of divided elements. After starting with the initial value of the design parameter, the average magnetic field inside the gap is calculated from the finite element solution of the magnetic circuit. Once the yield stresses of the MR fluid inside the gap are obtained from the magnetic field, the objective function can be calculated by Equation (8.11). Before starting the optimization procedure, the initial values for the design parameters of t, w, and dh are assigned to be 0.004 m, 0.01 m, and 0.005 m, respectively. Figure 8.6 shows the FEM that is analyzed using a commercial tool. Figure 8.6(a) is the result before optimization of the MR fan clutch and Figure 8.6(b) is the final result of optimization. As shown in Figure 8.6, the width of the coil is shorter than the initial condition, while the depth of the coil is longer than the initial condition.

Figure 8.7 shows the optimal solution of the MR fan clutch in a specific spatial limitation when a current of 1.2 A is applied to the coil. Figure 8.7 indicates that the objective function is converged to the constant value after the tenth iteration. As shown from Figure 8.7(a), the value of optimization results

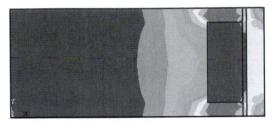

(a) Initial

(b) Optimization

FIGURE 8.6
Magnetic flux of the element material model. (From Kim, E.S. et al., *Smart Materials and Structures*, 19, 10, 2010. With permission.)

is minimized from 0.097 to 0.034 and thus, the final torque is increased to 29.32 Nm. It is also seen from Figure 8.7(b) that the effective length of the disk except the magnetic coil (*t*), the depth of the magnetic coil (*w*), and the width of the housing (*dh*) is changed to 0.0129 m, 0.0117 m, and 0.0854 m, respectively. Figure 8.8 shows a two-dimensional magnetic flux line and it can be observed that the magnetic flux lines are concentrated in the two paths that are located on the upper side and lower side of the magnetic coil. Using the design parameters that are optimized by FEM, the MR fan clutch is manufactured. The specific values of design parameters are given in Table 8.1. By considering the manufacturing environment, the effective length of the disk except the magnetic coil (*t*), the depth of the magnetic coil (*w*), and the width of the housing (*dh*) are finally chosen by 0.0130 m, 0.02 m, and 0.085 m, respectively. The vertical gap size (g_v) is determined to be 0.001 m and the horizontal gap size (g_h) is determined to be 0.003 m. The number of the gap is 2 because the disk of the MR fan clutch is single type. When the applied current is 1.2 A, the maximum yield shear stress τ_y is 49.2 kPa and the maximum controllable torque T_c is 29.3 Nm.

Figure 8.9 shows a photograph of an MR fan clutch that is manufactured based on the determined design parameters. Figure 8.9(a) is an assembly view of the MR fan clutch and Figure 8.9(b) is a component that consists of the housing, input shaft, and disk. The input shaft and the housing are designed to one part, and the output shaft is connected to the disk of the MR

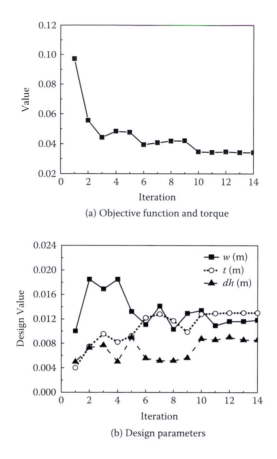

(a) Objective function and torque

(b) Design parameters

FIGURE 8.7
Optimal design results of the MR fan clutch. (From Kim, E.S. et al., *Smart Materials and Structures,* 19, 10, 2010. With permission.)

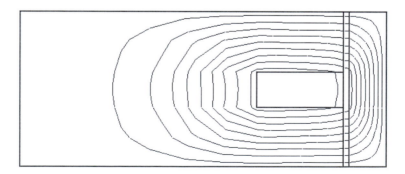

FIGURE 8.8
Two-dimensional magnetic flux lines of the optimized MR fan clutch. (From Kim, E.S. et al., *Smart Materials and Structures,* 19, 10, 2010. With permission.)

TABLE 8.1

Design Parameters of the MR Fan
Clutch

Design Parameter	Value
Height of disk, t	0.013 m
Depth of coils, w	0.012 m
Width of housing, dh	0.0085 m
Vertical gap size, g_v	0.001 m
Horizontal gap size, g_h	0.003 m
Max. electrical input	1.2 A
Max. field	50 A/mm
Max. yield shear stress, τ_y	49.2 kPa
Max. controllable torque	29.3 Nm

(a) MR fan clutch device assembly

(b) Disk of MR fan clutch

FIGURE 8.9
Photograph of the MR fan clutch.

fan clutch. The output shaft of the MR fan clutch is connected to the cooling fan and the input shaft is connected to the motor, which plays a role in transmitting the power of the MR fan clutch. As shown in Figure 8.9(b), aluminum is attached to the upper and lower sides of the disk to the concentrated magnetic flux. The magnetic coil is copper and the magnetic coil is fully wired in the area of the coil. The component material of the MR fan clutch is steel except for the upper and lower sides of the disk. In addition, a ball bearing is used to transmit power, and a retainer is used to seal leakage of MR fluid.

8.2.2 Controller Formulation

The main purpose is to realize a robust control system that can constantly maintain the desired coolant temperature of the vehicle engine system. Figure 8.10 is a block diagram of the engine cooling system for an automobile. As shown in Figure 8.10, the control system of the engine cooling system is composed of an MR fan clutch and associated controller. The input power is transmitted via DC motor and the MR fan clutch transmits the proper torque to the cooling fan, and finally, the cooling fan generates airflow. In this engine cooling system, the coolant is stored in the radiator and circulated by the hydraulic pump. The input and output temperatures of circulating coolant are sensed by thermocouples. After that, the sensed temperature is transmitted to the MR fan clutch controller. Then the control input (voltage) is calculated by the controller to be applied. The

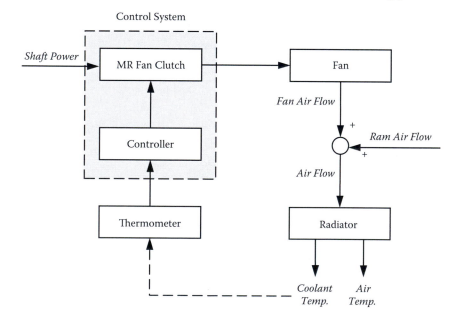

FIGURE 8.10
Temperature control block diagram of engine cooling system.

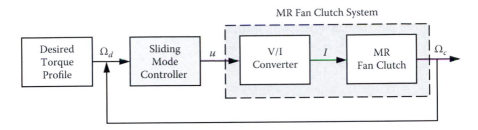

FIGURE 8.11
Control block diagram for the MR fan clutch system.

current is applied to the MR fan clutch via V/I (voltage/current) converter. Finally, the cooling fan is operated and the airflow is made. By these over-all sequences, the temperature of the engine cooling system is maintained as desired. It is remarked here that the state of vehicle is assumed idle in this experimental work. The control block diagram for the MR fan clutch system is shown in Figure 8.11. The MR fan clutch control system consists of the V/I converter and the MR fan clutch. The V/I converter and the MR fan clutch are modeled as the first-order system from experimental obser-vation (refer to Figure 8.14). Therefore, the overall MR fan clutch system can be considered to be the second-order system, which is multiplied by the V/I converter and the MR fan clutch serially. Figure 8.12 shows the transfer function from the input voltage to the output angular velocity of the cooling fan. Thus, the equation of the MR fan clutch system can be expressed as:

$$G(s) = \frac{\Omega(s)}{U(s)} = G_c(s)G_p(s) = \frac{1}{(t_c s + 1)(t_p s + 1)} \tag{8.13}$$

In the above, two independent systems, the V/I converter and the MR fan clutch, are multiplied serially as mentioned. $G(s)$ is the transfer function of the MR fan clutch system, $G_c(s)$ is the transfer function of the MR fan clutch, and $G_p(s)$ is the transfer function of the V/I converter. t_p is the time constant of the V/I converter and t_c is the time constant of the MR fan clutch. t_p and t_c can be experimentally determined from the step response (refer to Figure 8.14). In Equation (8.13), $\Omega(s)$ is the angular velocity of the cooling fan and $U(s)$ is the control input of the control system, which physically means voltage level.

FIGURE 8.12
Transfer function between input voltage and output angular velocity.

Once it is assumed that the initial condition of Equation (8.13) is zero, the equation of the angular velocity control system can be defined by

$$t_p t_c \ddot{\Omega}(t) + (t_p + t_c)\dot{\Omega}(t) = u(t) + d(t) \tag{8.14}$$

$$u(t) = \kappa V(t) \tag{8.15}$$

In the above, $u(t)$ is the input voltage of the V/I converter and κ is a voltage gain that is calculated by a controller. $d(t)$ is the external disturbance of the MR fan clutch system. The state space model can be derived from Equation (8.14) as:

$$X(t) = \begin{bmatrix} x_1 & x_2 \end{bmatrix}^T = \begin{bmatrix} \Omega & \dot{\Omega} \end{bmatrix}^T \tag{8.16}$$

$$\dot{X}(t) = \mathbf{A}X(t) + \mathbf{B}u(t) + \mathbf{D}d(t) \tag{8.17}$$

where

$$\mathbf{A} = \begin{bmatrix} 0 & 1 \\ -\dfrac{1}{t_p t_c} & -\dfrac{t_p + t_c}{t_p t_c} \end{bmatrix}, \ \mathbf{B} = \begin{bmatrix} 0 \\ \dfrac{1}{t_p t_c} \end{bmatrix}, \ \mathbf{D} = \begin{bmatrix} 0 \\ \dfrac{1}{t_p t_c} \end{bmatrix} \tag{8.18}$$

In the above, x_i is the state variable, $\mathbf{A}$ is the control system matrix, $\mathbf{B}$ is the control input matrix, $\mathbf{D}$ is the disturbance matrix, and u is the control input of the system.

In order to guarantee the stability of the cooling control system, a robust sliding mode controller [17, 18] is designed in this section. The control variable is the angular velocity of the MR fan clutch system. The angular velocity is directly related to the temperature of the coolant circulated in the engine cooling system. This relation is experimentally identified in this subsection (refer to Figure 8.15). Thus, the control tracking error is defined as:

$$e(t) = X(t) - X_d(t) = \begin{bmatrix} x_1 - x_{d1} & x_2 - x_{d2} \end{bmatrix} \tag{8.19}$$

As a next step, using the tracking error, the sliding surface is defined as:

$$S(e(t)) = \left(\frac{d}{dt} + c\right) \cdot e_1(t) + e_2(t) = Ce(t), \ c > 0 \tag{8.20}$$

where C is the vector that denotes a gradient of the sliding surface and c is the sliding surface coefficient to be determined so that the sliding surface itself is stable. In order to drive the system state to the sliding surface, $S(t) = 0$, a control law has to be properly designed such that the following sliding mode condition is satisfied:

$$S(e(t)) \cdot \dot{S}(e(t)) < 0 \tag{8.21}$$

In order to satisfy the above condition, the control input is determined by

$$u = -(CB)^{-1}(CAX(t) - C\dot{X}_d(t)) - k\,\text{sgn}(S(t)), \quad k > |d(t)| \qquad (8.22)$$

In the above, k is the discontinuous control gain. It is noted here that the uncertainty of time constants (V/I converter time constant, t_p, and MR fan clutch time constant, t_s) is considered as:

$$\hat{t}_p = t_p + \Delta t_p, \quad |\Delta t_p| \le a$$
$$\hat{t}_s = t_s + \Delta t_s, \quad |\Delta t_s| \le b \qquad (8.23)$$

In the above, Δt_p and Δt_s is the variation of each time constant and a and b are uncertain parameters whose limitations are known. Because the discontinuous controller in Equation (8.22) causes undesirable chattering [18], the replacement of the discontinuous controller by the continuous controller using the saturation function is needed as:

$$u = -(CB)^{-1}(CAX(t) - C\dot{X}_d(t)) - k\,sat(S(t)), \quad k > |d(t)| \qquad (8.24)$$

$$sat(S(t)) = \begin{cases} S(t)/\varepsilon, & |S(t)| \le \varepsilon \\ \text{sgn}(S(t)), & |S(t)| > \varepsilon \end{cases} \qquad (8.25)$$

In the above, ε is the boundary layer width of the saturation function, which adjusts the magnitude of the chattering.

8.2.3 Experimental Results

In order to evaluate performance of the control system associated with the sliding mode controller, an experimental apparatus and instrumentation is established as shown in Figure 8.13. The input power is generated by a DC motor, and the current level according to the input-voltage is measured using dSPACE (A/D converter and D/A converter). Before implementing the temperature controller, the step response of the V/I converter and the MR fan clutch is measured. Figure 8.14 shows the step response of the V/I converter and the MR fan clutch. Based on the step response, the time constant can be determined. In this test, the following nominal value of the time constant is used: $t_p = 162$ ms, $t_c = 337$ ms. The variation of each time constant is assumed to be ±10% of the nominal value. This variation is considered to determine proper control gain k in Equation (8.24). Figure 8.15 shows the temperature target of the MR fan clutch according to the fan speed. Based on the relationship between the temperature and the cooling fan speed, the sliding mode control logic is realized using MATLAB Simulink®.

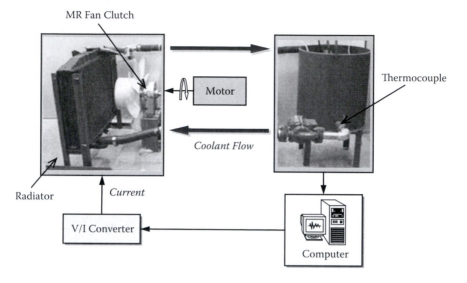

FIGURE 8.13
Experimental apparatus for temperature control using an MR fan clutch.

In this test, the desired temperature is set to be 45 ± 1°C. Before activating the sliding mode controller, the cooling system is operated without control input to the MR fan clutch. Figure 8.16 shows the coolant temperature in the condition without control action. In this case, the cooling fan is transmitted to the torque only by the viscosity of the MR fluid. Thus, the coolant temperature cannot reach the desired temperature. Figure 8.17 shows the coolant temperature control result using the sliding mode controller. Figure 8.17(a) is the temperature control response and Figure 8.17(b) is the regulating error. As shown in Figure 8.17(a), the coolant temperature reaches the desired steady state value 45°C after 50 s, and after that, the coolant temperature is maintained at the desired value of 45°C. The regulating error is below ±0.5°C. In this test, a typical PID (proportional–integral–derivative) controller is also experimentally implemented in order to compare with the control performance of the sliding mode controller. The desired temperature target is the same (45 ± 1°C) and all environmental conditions are the same as for the sliding mode controller. The control gain of the PID controller is determined so that maximum control input should be the same as that of the sliding mode controller. The experimental result of the temperature control using the PID controller is shown in Figure 8.18. As shown in Figure 8.18(a), as soon as the current input is applied to the MR fluid domain, the temperature started to decrease and 100 s later, the desired temperature was reached. It is clearly seen from Figure 8.17 and Figure 8.18 that the sliding mode controller well follows the desired temperature with faster reaching time than the PID controller under the same experimental conditions.

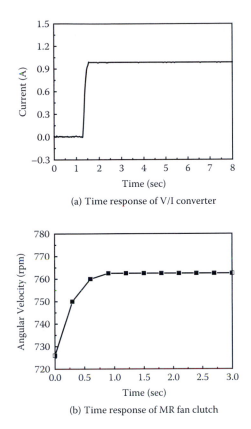

(a) Time response of V/I converter

(b) Time response of MR fan clutch

FIGURE 8.14
Time response of the MR fan clutch system. (From Kim, E.S. et al., *Smart Materials and Structures,* 19, 10, 2010. With permission.)

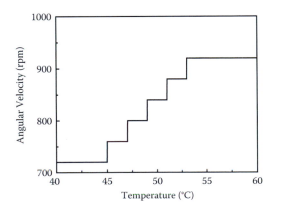

FIGURE 8.15
The relationship between angular velocity and temperature. (From Kim, E.S. et al., *Smart Materials and Structures,* 19, 10, 2010. With permission.)

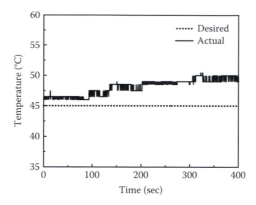

FIGURE 8.16
The coolant temperature without control action. (From Kim, E.S. et al., *Smart Materials and Structures*, 19, 10, 2010. With permission.)

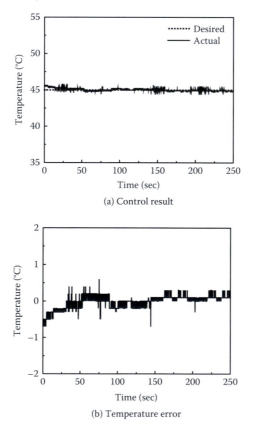

(a) Control result

(b) Temperature error

FIGURE 8.17
Temperature control response using the sliding mode controller. (From Kim, E.S. et al., *Smart Materials and Structures*, 19, 10, 2010. With permission.)

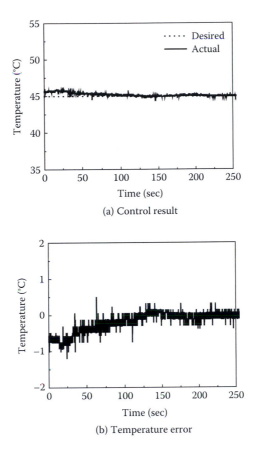

(a) Control result

(b) Temperature error

FIGURE 8.18
Temperature control response using the PID controller. (From Kim, E.S. et al., *Smart Materials and Structures*, 19, 10, 2010. With permission.)

8.3 MR Seat Damper

8.3.1 Damper Design

A cylindrical-type MR seat damper is designed based on the damping force level of a conventional seat damper for a commercial truck. The MR product (MRF-132LA) from the Lord Corporation has been employed as a controllable fluid. When the magnetic field of H is applied to the MR fluid, the yield stress (τ_y) can be expressed by

$$\tau_y(H) = \alpha_m H^{\beta_m} \tag{8.26}$$

where α_m and β_m are intrinsic values of MR fluid to be experimentally determined. The field-dependent yield stress given by Equation (8.26) is

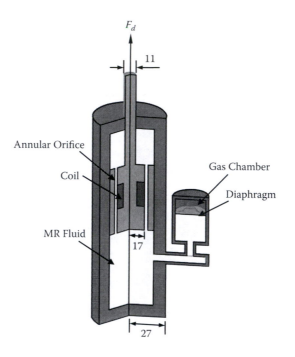

FIGURE 8.19
Schematic diagram of the proposed MR seat damper.

to be incorporated with the damping force model of the proposed MR seat damper. Figure 8.19 presents the schematic configuration of the MR seat damper. The piston head with magnetic cores is moving up and down, and the gas chamber is located outside as an accumulator of the MR fluid. In order to simplify the analysis of the MR seat damper, it is assumed that the MR fluid is incompressible and fluid inertia is assumed to be negligible. For laminar flow in the annular orifice, the flow resistance under the zero magnetic field is given by

$$R_m = \frac{12\eta_1 L_m}{b_1 h_m^3} \tag{8.27}$$

where L_m is the orifice length, b_1 is the orifice width, h_m is the orifice gap, and η_1 is the viscosity of the MR fluid. On the other hand, the pressure drop due to the magnetic field is given by

$$\Delta P_{MR} = 2\frac{L_m}{h_m}\tau_y(H) = 2\frac{L_m}{h_m}\alpha_m H^{\beta_m} \tag{8.28}$$

where $H = NI/2h_m$. Here, N is the number of coil turns and I is the input current.

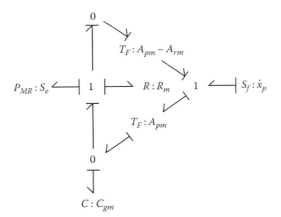

FIGURE 8.20
Bond graph model of the proposed MR seat damper.

Now, from the bond graph model in Figure 8.20, the damping force of the MR damper is obtained by neglecting the effect of the fluid inertia and frictional force as:

$$F_d = \frac{A_{rm}^2}{C_{gm}} x_p + (A_{pm} - A_{rm})^2 R_m \dot{x}_p + (A_{pm} - A_{rm})\Delta P_{MR}\,\text{sgn}(\dot{x}_p) \qquad (8.29)$$

where C_{gm} is the gas compliance, A_{rm} is the piston-rod area, A_{pm} is the piston area, and sgn($\cdot$) is the signum function. It is noted that the third term is controllable damping force by the intensity of the magnetic field. The damping force given by Equation (8.29) can be rewritten as

$$F_d = k_e x_p + c_e \dot{x}_p + F_{MR} \qquad (8.30)$$

where k_e is the effective stiffness due to the gas pressure and c_e is the effective damping coefficient due to the viscosity of the MR damper. The size and level of required damping force adopted in this test are chosen based on a conventional passive oil seat damper for a commercial truck. The damping force given by Equation (8.30) is evaluated with respect to the magnetic field in order to determine principal design parameters. The designed MR seat damper has the following specifications: orifice gap: 1.8 mm, orifice length: 44 mm, diameter of the piston head: 46 mm, number of coil turns: 60, and diameter of the coil (copper wire): 0.75 mm.

Figure 8.21 presents the measured damping force characteristics of the MR seat damper. Figure 8.21(a) is obtained by exciting the MR seat damper with a frequency of 1.6 Hz and an amplitude of ±10 mm. It is identified that the damping force of 173 N in the absence of the input field is increased up to 421 N by applying the input current of 2 A. Figure 8.21(b) is obtained by increasing excitation frequency from 0.6 Hz to 4.0 Hz, where the excitation

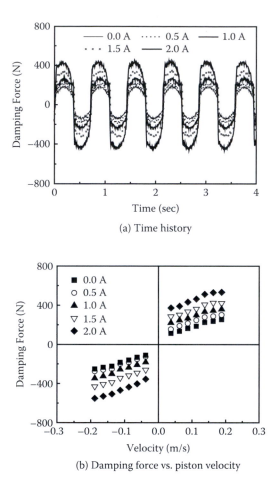

FIGURE 8.21
Damping force characteristics of the proposed MR seat damper. (From Choi, S.B. et al., *International Journal of Vehicle Design*, 31, 2, 2003. With permission.)

amplitude is maintained to be constant at 20 mm (±10 mm). It is clearly observed that the damping force increases as the input current increases. This directly indicates that a certain desired damping force can be achieved by just controlling the input current regardless of the piston velocity.

8.3.2 System Modeling

A full-vehicle model consisting of passive primary dampers and semi-active MR seat dampers is shown in Figure 8.22. The excitation input from the road is transmitted through to the cabin floor. This then causes unwanted vibrations on the driver. For the simplification of dynamic modeling, only the straight-going motion of the vehicle is considered. Accordingly, it is assumed

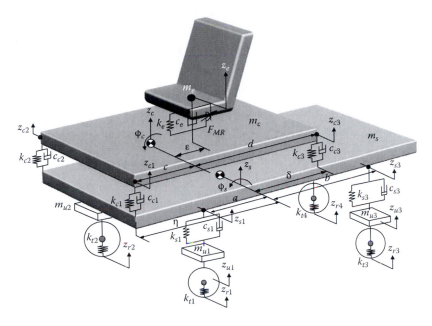

FIGURE 8.22
Mechanical model of the full-vehicle suspension system.

that only the vertical and the pitching motion of the vehicle exist and the rolling motion is ignored in this study. Hence, the state variables are defined as

$$x = \left[z_{u1}, \dot{z}_{u1}, z_{u2}, \dot{z}_{u2}, z_{u3}, \dot{z}_{u3}, z_{u4}, \dot{z}_{u4}, z_s, \dot{z}_s, \phi_s, \dot{\phi}_s, z_c, \dot{z}_c, \phi_c, \dot{\phi}_c, z_e, \dot{z}_e \right]^T \quad (8.31)$$

Now, let us derive governing equations of motion for the full-vehicle model installed with passive primary and semi-active MR seat suspensions [15].

$$m_{u_i} \ddot{z}_{u_i} = f_{s_i} - f_{t_i}, \quad i = 1, 2, 3, 4$$

$$m_s \ddot{z}_s = -\sum_{i=1}^{4} f_{s_i} + \sum_{j=1}^{4} f_{c_j}$$

$$J_s \ddot{\phi}_s = a \sum_{i=1}^{2} f_{s_i} - b \sum_{i=3}^{4} f_{s_i} - \eta \sum_{j=1}^{2} f_{c_j} + \delta \sum_{j=3}^{4} f_{c_j}$$

$$m_c \ddot{z}_c = -\sum_{i=1}^{4} f_{c_i} + f_e + F_{MR}$$

$$(8.32)$$

$$J_c \ddot{\phi}_c = \varepsilon(f_e + F_{MR}) + c \sum_{j=1}^{2} f_{c_j} - d \sum_{j=3}^{4} f_{c_j}$$

$$m_e \ddot{z}_e = -f_e - F_{MR}$$

where

$$f_{s_i} = k_{s_i}(z_s - a\phi_s - z_{u_i}) + c_{s_i}(\dot{z}_s - a\dot{\phi}_s - \dot{z}_{u_i}), \quad i = 1,2$$

$$f_{s_i} = k_{s_i}(z_s + b\phi_s - z_{u_i}) + c_{s_i}(\dot{z}_s + b\dot{\phi}_s - \dot{z}_{u_i}), \quad i = 3,4$$

$$f_{c_i} = k_{c_i}\left[(z_c - c\phi_c) - (z_s - \eta\phi_s)\right] + c_{c_i}\left[(\dot{z}_c - c\dot{\phi}_c) - (\dot{z}_s - \eta\dot{\phi}_s)\right], \quad i = 1,2$$

$$f_{c_i} = k_{c_i}\left[(z_c + d\phi_c) - (z_s + \delta\phi_s)\right] + c_{c_i}\left[(\dot{z}_c + d\dot{\phi}_c) - (\dot{z}_s + \delta\dot{\phi}_s)\right], \quad i = 3,4 \qquad (8.33)$$

$$f_e = k_e(z_e - z_c - \varepsilon\phi_c) + c_e(\dot{z}_e - \dot{z}_c - \varepsilon\dot{\phi}_c)$$

$$f_{ti} = k_{ti}(z_{ui} - z_{ri}), \quad i = 1,2,3,4$$

It is noted that the cabin floor displacement is represented by $z_c + \varepsilon\phi_c$. This cabin floor input is to be applied to the MR seat suspension system via the hydraulic servo-actuator. It is also noted that $z_c + \varepsilon\phi_c$ is not the ideal displacement source for the seat suspension. It is incorporated with the dynamic model of the full-vehicle system given by Equation (8.32). All variables used in Equation (8.32) and Equation (8.33) are defined in Table 8.2 with specific values.

TABLE 8.2

System Parameters of the Full-Vehicle Model

Parameter	Value	Parameter	Value
Unsprung mass 1 ($m_{u1,2}$)	803 kg	Primary suspension damping (c_{si})	33,000 Ns/m
Unsprung mass 2 ($m_{u3,4}$)	1503 kg	Cabin suspension damping (c_{ci})	5073.5 Ns/m
Car body mass (m_s)	3450 kg	Seat suspension damping (c_e)	750 Ns/m
Cabin mass (m_c)	950 kg	From car C.G. to steer axle (a)	1.15 m
Seat and driver mass (m_e)	75 kg	From car C.G. to rear steer axle (b)	2.8 m
Car body inertia (J_s)	9500 kg·m²	From car C.G. to front cabin suspension (η)	1.8 m
Cabin inertia (J_c)	800 kg·m²	From car C.G. to rear cabin suspension (δ)	0.2 m
Tire stiffness (k_{ti})	4000 kN/m	From cab C.G. to cabin front end (c)	0.85 m
Primary suspension stiffness (k_{si})	400 kN/m	From cab C.G. to cabin rear end (d)	1.15 m
Cabin suspension stiffness (k_{ci})	63757.5 N/m	From cab C.G. to seat suspension (ε)	0.01 m
Seat suspension stiffness (ke)	4900 N/m		

Among many potential candidates for control algorithms, the semi-active skyhook controller [19] is useful for the suspension system. It is well known that the logic of the skyhook is simple and easy to implement in a practical field. The desired damping force for an MR seat damper is set by

$$u = C_{sky\_MR} \cdot \dot{z}_e = F_{MR} \tag{8.34}$$

where C_{sky_MR} is the gain of the skyhook controller. The control gain physically indicates the damping coefficient. On the other hand, the damping force of the seat suspension system needs to be controlled depending upon the motion of the suspension travel. Therefore, the following actuating condition is normally imposed.

$$u = \begin{bmatrix} F_{MR}, & for \ F_{MR} \cdot (\dot{z}_e - \dot{z}_c - \varepsilon \dot{\phi}_c) > 0 \\ 0, & for \ F_{MR} \cdot (\dot{z}_e - \dot{z}_c - \varepsilon \dot{\phi}_c) \le 0 \end{bmatrix} \tag{8.35}$$

The above actuating conditions physically imply that the activating of the controller only assures the increment of energy dissipation of the stable system. Once the control input is determined, the input current to be applied to the MR seat damper is, subsequently, obtained as [15]:

$$I = \frac{2h_m}{N} \left[u \cdot \frac{h_m}{8\alpha_m \cdot L_m (A_{pm} - A_{rm})} \right]^{1/\beta_m} \tag{8.36}$$

8.3.3 Vibration Control Results

From a practical point of view, the passive primary dampers and semi-active MR seat damper should be installed on the full-vehicle and its effectiveness for the ride quality needs to be tested under various road conditions. However, it is very expensive to completely build the entire full-vehicle system with hardware from the start. Therefore, we adopt an HILS in which a full-vehicle system model consisting of the conventional passive primary and cabin suspensions is incorporated with actual hardware of the proposed semi-active MR seat suspension. Figure 8.23 presents a schematic configuration of the HILS method. In the real-time simulation, the road and driving conditions are adopted for the conventional cab-over-engine (COE)-type commercial vehicle model.

The full-vehicle model of a commercial vehicle solved by a digital computer in real-time is to be excited from two types of road profiles. The first profile, shown in Figure 8.24 and normally used to reveal the transient response characteristic, is a bump described by

$$z_{ri} = z_b [1 - \cos(w_r t)], \qquad\qquad i = 1, 2$$
$$z_{rj} = z_b [1 - \cos(w_r (t - D_{car} / V_s))], \qquad j = 3, 4 \tag{8.37}$$

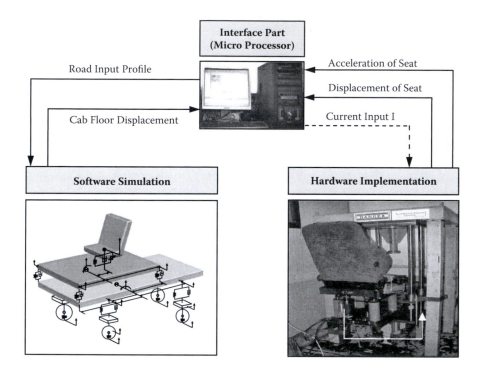

FIGURE 8.23
Schematic configuration of the HILS.

where $w_r = 2\pi V_s/D$, z_b (small-sized bump = 0.02 m, large-sized bump = 0.04 m) is half of the bump height, $D (= 0.75$ m) is the width of the bump, $D_{car} (= 2.4$ m) is the wheel base, which is defined as the distance between the front wheel and the rear one, and V_s is the vehicle velocity. In the bump excitation, the vehicle travels the bump with a constant velocity of 3.08 km/h (= 0.856 m/s). The second type of road profile, normally used to evaluate the frequency response, is a stationary random process [20] with zero mean described by

$$\dot{z}_{ri} + \rho_r V_s z_{ri} = V_s W_{ni}, \quad i = 1, 2, 3, 4 \tag{8.38}$$

where W_{ni} is the white noise with intensity $2\sigma^2 \rho_r V_s$, $\rho_r (= 0.45 \text{ m}^{-1})$ is the road roughness parameter, and $\sigma^2 (= 300 \text{ mm}^2)$ is the covariance of road irregularity. In the random excitation, the vehicle travels on the paved road with constant velocity of 72 km/h (= 20 m/s).

The system parameters of the full-vehicle model are adopted based on a commercial heavy truck in this test, and are listed in Table 8.2. The cabin floor displacement is calculated and then the displacement signal is converted to the hydraulic control unit for controlling the road input to the platform. When the vibration is transmitted to the MR seat suspension, the dynamic

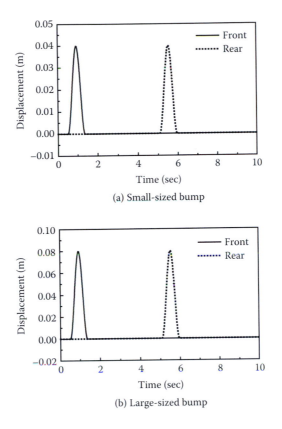

FIGURE 8.24
Bump road profiles for the HILS. (From Choi, S.B. et al., *International Journal of Vehicle Design*, 31, 2, 2003. With permission.)

response signals, which are acquired from the linear variable differential transformer (LVDT) and accelerometer, are converted to digital signals via the A/D converter, which has 12 bits.

Figure 8.25 and Figure 8.26 present the small- and large-sized bump control responses of the full-vehicle system. As clearly observed from the results, both vertical displacement and acceleration at the driver's seat have been significantly reduced under both small- and large-sized bump excitations by controlling the semi-active MR seat suspension. Especially in the large-sized bump response, the displacement and acceleration are reduced up to 35% and 20%, respectively, by controlling the MR seat suspension. It is also seen from Figure 8.25(c) and Figure 8.26(c) that input currents are supplied to the MR seat damper according to the semi-active control conditions. Figure 8.27 shows the random road control response of the full-vehicle system. It is obvious that the PSD of vertical acceleration at the driver's seat is considerably attenuated by activating the MR seat damper. The maximum magnitude is reduced by 40% at the resonant frequency.

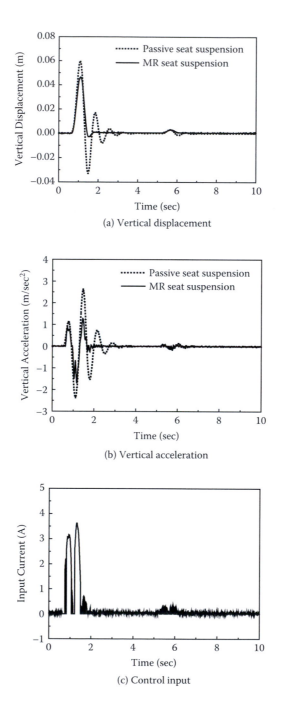

FIGURE 8.25
Control responses of the full-vehicle system under small-sized bump excitation. (From Choi, S.B. et al., *International Journal of Vehicle Design*, 31, 2, 2003. With permission.)

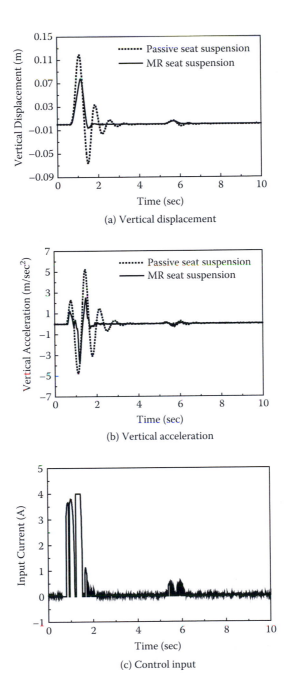

FIGURE 8.26

Control responses of the full-vehicle system under large-sized bump excitation. (From Choi, S.B. et al., *International Journal of Vehicle Design*, 31, 2, 2003. With permission.)

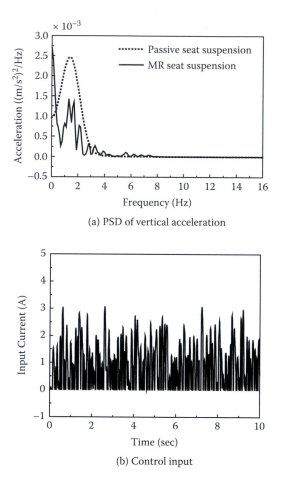

FIGURE 8.27

Control responses of the full-vehicle system under random excitation. (From Choi, S.B. et al., *International Journal of Vehicle Design*, 31, 2, 2003. With permission.)

8.4 Some Final Thoughts

In this chapter, two MR applications were introduced for heavy vehicles. One is the fan clutch for an automotive engine cooling system. A drum type of the MR clutch was devised and its principal design parameters were determined using FEA under consideration of spatial limitations and dimensions of the conventional fan clutch. The field-dependent angular velocity and response time to step input were measured for the manufactured MR fan clutch. After constructing an experimental apparatus, which can reflect a real engine cooling system, the sliding mode controller was implemented to adjust angular velocity of the fan according to the desired temperature. It has been clearly identified that the proposed sliding mode controller well follows the desired temperature with a small regulating error. In addition, it was shown that the reaching time of the sliding mode controller is faster than that of the PID controller, owing to the robustness against the uncertain time constant of the converter and fan clutch. It is generally well known that an automotive fan clutch is one of the major equipments in an engine cooling system and the performance of the clutch is directly related to fuel efficiency of the engine. Therefore, it is expected that the proposed control system can be effectively utilized for the automotive cooling system to improve the fuel efficiency.

The other is the damper for a seat suspension system. Vibration control responses of a heavy vehicle equipped with the passive primary dampers and a semi-active MR seat damper have been evaluated under both bump and random road conditions. It has been demonstrated via HILS method that both vertical displacement and acceleration at the driver's seat have been remarkably attenuated under bump road conditions by activating the MR seat damper associated with the semi-active skyhook controller. In addition, the power spectrum density of the acceleration under the random road excitation was significantly reduced by controlling the MR seat damper.

In the field of heavy industrial vehicles, the MR applications have a great potential to overcome severe working environments and provide good vehicle performance and ride comfort. It is remarked that control robustness and durability of the MR seat damper needs to be further investigated further for practical application.

References

[1] Queslati, F. and Sankar, S. 1992. Performance of a fail-safe active suspension with limited state feedback for improved ride quality and reduced pavement loading in heavy vehicles. *SAE Paper* 922474: 796–804.

[2] Kim, E. S., Choi, S. B., Park, Y. G. and Lee, S. 2010. Temperature control of automotive engine cooling system utilizing MR fan clutch. *Smart Materials and Structures* 19: 1–10.

[3] Neelakantan, V. A. and Washington, G. N. 2005. Modeling and reduction of centrifuging in magnetorheological (MR) transmission clutches for automotive applications. *Journal of Intelligent Material Systems and Structures* 16: 703–712.

[4] Kavlicoglu, N. C., Kavlicoglu, B. M., Liu, Y., Evrensel, C. A., Fuchs, A., Korol, G., and Gordaninejad, F. 2007. Response time and performance of a high-torque magnetorheological fluid limited slip differential clutch. *Smart Materials and Structures* 16: 149–159.

[5] Hakogi, H., Ohaba, M., Kuramochi, N., and Yano, H. 2005. Torque control of a rehabilitation teaching robot using magnetorheological fluid clutches. *JSME International Journal* 48: 501–507.

[6] Choi, S. B., Hong, S. R., Cheong, C. C., and Park, Y. K. 1999. Comparison of field-controlled characteristics between ER and MR clutches. *Journal of Intelligent Material Systems and Structures* 10: 615–619.

[7] Benetti, M. and Dragoni, E. 2006. Nonlinear magnetic analysis of multi-plate magnetorheological brakes and clutches. *Proceedings of the COMSOL Users Conference.* 1–5.

[8] Choi, S. B. and Han, Y. M. MR seat suspension for vibration control of a commercial vehicle. *International Journal of Vehicle Design* 31: 201–215.

[9] Weiss, K. D., Duclos, T. G., Carlson, J. D., Chrzan, M. J., and Margida, A. J. 1993. High strength magneto- and electro-rheological fluids. *SAE Paper* 932451.

[10] Dyke, S. J., Spencer, Jr., B. F., Sain, M. K., and Carlson, J. D. 1996. Seismic response reduction using magnetorheological damper. *Proceedings of IFAC World Congress* 1: 145–150.

[11] Carlson, J. D. and Spencer, Jr., B. F. 1996. Magnetorheological fluid dampers for semi-active seismic control. *Proceedings of the 3rd International Conference on Motion and Vibration Control* 2: 35–40.

[12] Pang, L., Kamath, H. M., and Wereley, N. M. 1998. Dynamic characterization and analysis of magnetorheological damper behavior. *Proceedings of the 5th SPIE Symposium on Smart Materials and Structure, Passive Damping and Isolation* 3325–3327.

[13] Carlson, J. D. and Weiss, K. D. 1994. A growing attraction to magnetic fluids. *Machine Design* 8: 61–66.

[14] Choi, S. B., Nam, M. H., and Lee, B. K. 2000. Vibration control of a MR seat damper for commercial vehicles. *Journal of Intelligent Material Systems and Structures* 11: 936–944

[15] Nam, M. H. 2001. Performance evaluation of vehicle suspension system featuring MR seat damper. Ph. D. dissertation, Inha University, Incheon, Korea.

[16] Jolly, M. R., Bender, J. W., and Carlson, J. D. 1997. Properties and applications of commercial magnetorheological fluids. *The International Society for Optical Engineering* 3327: 262–275.

[17] Slotine, J. J. E. and Sastry, S. S. 1994. Tracking control of nonlinear systems using sliding surfaces with application to robot manipulators. *International Journal of Control* 38: 465–492.

[18] Choi, S. B., Park, D. W., and Jayasuriya, S. 1994. A time-varying sliding surface for fast and robust tracking control of second-order uncertain systems. *Automatica,* 30: 809–904.

[19] Karnopp, D., Crosby, M. J., and Harwood, R. A. 1995. Vibration control using semi-active force generators. *Journal of Engineering for Industry* 96: 619–629.

[20] Nigam, N. C. and Narayanan, S. 1994. *Applications of Random Vibrations,* Berlin: Springer-Verlag.

9

Haptic Applications for Vehicles

9.1 Introduction

Haptic technology is based on the sense of touch. It provides a new communication method that closely resembles the subtle, non-visual, and non-auditory communication methods found among biological beings. The mechanical interface device providing stimulus information such as tactile sensation and kinesthetic force to a user is called a haptic device or haptic interface. Goertz developed the first force feedback system during the 1940s for a master-slave manipulator designed to handle radioactive substances. Interfaces with the human sense of touch have actively been researched lately in various application fields such as automotive and robotic to achieve great realism and interactivity including good accuracy and high performance [1, 2]. In the automotive field in particular, much attention is being placed upon the areas of driving performance, safety, driver comfort, and convenience. Haptic technology holds the key to enhancing capabilities in these areas.

Until now, various haptic devices have been developed for automotive vehicles. BMW cars now use the iDrive user interface to control most secondary vehicle systems [2]. Bengtsson, Grane, and Isaksson proposed the use of a haptic interface for in-vehicle comfort functions and evaluated it in an experimental study [3]. Aoki and Murakami developed a haptic pedal to provide feedback of the estimated road conditions to drivers [4]. Kobayashi et al. proposed an accelerator pedal device that provides haptic feedback to avoid rear-end collision situations [5]. However, in most of the previous works, conventional actuators such as electric motors have been employed. On the other hand, the conventional actuators have complex actuation mechanisms and struggle to provide continuous and precise force control. More advanced haptic devices need to be further developed because the human sense of touch is far more sensitive than that of sight or hearing.

Therefore, several studies on haptics have been recently conducted through the adoption of smart materials such as magnetorheological (MR) fluids, electrorheological (ER) fluids, and piezoelectric materials. Among them, MR fluids are actively researched as an actuating fluid for various applications [6–8]. In particular, the MR clutch or brake mechanism featuring continuously controllable torque has been widely studied due to its benefits including high stability and performance [9, 10]. As mentioned in Chapter 1, the phenomenological behavior of MR fluid can make the resistance of external forces or pressures possible and offers several benefits such as good stability, reliable control performance, and compact design. These are ideal capabilities for haptic systems. The control of rheological properties by a magnetic field provides the possibility for MR haptic devices to accurately mirror remote or virtual compliance. The advantages of MR fluids have led to several studies for haptic devices. Scilingo et al. [11] proposed a haptic display based on MR fluids to provide a pinch grasp. Li developed a haptic device that uses MR fluids [12]. In addition, various haptic devices based on MR fluids have been suggested for automotive systems to enhance performance and develop new functions.

In this chapter, we focus on two fascinating haptic applications utilizing MR fluids for passenger vehicles: the multi-functional control knob for vehicle instrument panels and the haptic cue accelerator for manual transmission vehicles.

Section 9.2 introduces a multi-functional control knob, an ideal candidate for the consolidation of various vehicular instrument controls [13]. In recent years, vehicles provide various comfort functions to drivers such as audio, power windows, road and traffic information, and air conditioning. The growing number of comfort functions requires several different vehicular instrument control devices to be installed on a dashboard as secondary vehicle controls. These various secondary control devices can decrease a driver's attention, one of the most common causes of traffic accidents. However, the combination of several in-vehicle functions into a single device can enable ride information to be transmitted without requiring a driver's visual attention. This can be achieved by the adoption of a multi-functional control knob based on a driver–vehicle haptic interaction. An enhanced haptic actuator using MR fluid can realize both rotary motions and push motions of in-vehicle functions with precise tactile feeling and sufficient stopping force.

As a first step, a haptic mechanism capable of both rotary and push motions is devised, and design parameters are optimally determined to maximize a relative control torque via finite element analysis (ANSYS parametric design language). The haptic mechanism is then established within a single knob. The governing torque and force motions of haptic knobs, which are capable of both rotary and push motions, are mathematically modeled and experimentally verified. In-vehicle comfort functions are also constructed in a virtual environment that communicates with the manufactured haptic knob. The haptic device is then incorporated with the in-vehicle function in the virtual environment. Subsequently, a feed-forward controller is synthesized based on position information and torque/force map for each in-vehicle function. Control performances such as the reflection force of the haptic device are experimentally evaluated and demonstrated.

Section 9.3 presents a concept of a haptic cue for manual vehicle transmissions, a "haptic cue accelerator" featuring MR fluids [14,15]. Over the past few years, various driver supportive systems have been developed to increase driving comfort and safety as well as traffic efficiency. Fuel economy has received much emphasis recently in automobile engineering to reduce energy costs and global emissions. Therefore, many researchers have widely studied optimal gear shifting of vehicle transmissions to enhance fuel economy [16–18]. Most of these works are conducted with automatic transmissions such as continuously variable transmissions. However, in manual transmissions, research considering optimal gear shifting remains very limited because gear shifting highly depends on personal driving styles. Therefore, a new driver supportive scheme must be synthesized to allow drivers to recognize the optimal moment of gear shifting with minimum fuel consumption. This can be achieved by integrating haptic technology into acceleration mechanisms. This new driver supportive function can eventually reduce fuel consumption and vehicle emissions by enabling optimal gear shifting. This function can be realized by an accelerator pedal that has a haptic cue device.

The proposed system only transmits cue signals to allow the driver to determine the gear shifting timing and never interferes with the vehicle's accelerating function. Therefore, this system should be carefully designed to ensure that the haptic force on the accelerator never works against the driver's intention because one possible safety problem is that excessive haptic force can disturb a driver's manipulation of the accelerator.

As a first step, a haptic cue device is devised by adopting MR fluid and the brake mechanism rotary type while considering the spatial limitations of a mid-size passenger vehicle. After optimally determining design parameters by finite element analysis (FEA) using the ANSYS parametric design language, the MR haptic cue device is manufactured and integrated with a real accelerator to produce the MR haptic cue accelerator. In addition, a simple virtual vehicle emulating the engine operation of a passenger vehicle is constructed using dSPACE and MATLAB Simulink®, which communicates with the manufactured haptic device. After the field-dependent torque characteristics of the manufactured haptic cue accelerator are evaluated, the ensuing control performances are experimentally observed using the feed-forward haptic cue control strategy.

9.2 Multi-Functional MR Control Knob

9.2.1 Configuration

A cylindrical type mechanism is devised in order to achieve both rotary motion and push motion within a single device. Figure 9.1 shows a schematic configuration of the proposed multi-functional haptic knob. The operational knob is connected to a cylindrical electromagnetic disk, which has a flux guide and an electromagnetic coil, and it is assembled in housing

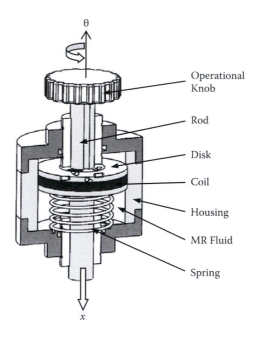

FIGURE 9.1
Schematic configurations of the proposed MR haptic knob.

with a specific gap fully filled with MR fluid. A coil spring is also installed below the electromagnetic disk to produce a restoring force at push motion. The MR product (MRF-132DG) from the Lord Corporation has been employed as a controllable fluid. Remembering the Bingham model of the MR fluid in Chapter 1, the shear stress (τ) under a magnetic field is expressed by

$$\tau = \tau_y(\cdot) + \eta \dot{\gamma} \tag{9.1}$$

where η is the dynamic viscosity, γ is the shear rate, and $\tau_y(\cdot)$ is the dynamic yield stress of the MR fluid. It is noted that the applied magnetic field can be expressed by one of magnetic flux density (B) and magnetic field strength (H), and they are used for design optimization and controller synthesis, respectively, in this section.

Among several operation modes of an MR fluid-based device, a shear mode is developed at the rotary motion of the MR knob. Therefore, the torque model is mathematically described as:

$$T = T_c + T_\eta + T_f$$

$$= 4\pi R^2 d \cdot \tau_y(\cdot) + \frac{4\pi \eta R^3 \dot{\theta} h}{g} + T_f \tag{9.2}$$

where $\dot{\theta}$ is the rotational velocity of the knob in rotary motion. R and h are the outer radius and length of the electromagnetic disk. g is the gap between the disk and housing. d is the half length of the electromagnetic disk except of width of the magnetic coil, which acts as a magnetic flux guide for MR fluids. As shown in the above equation, the produced torque (T) consists of the controllable torque T_c, viscous torque T_η, and friction torque T_f. The friction torque is mainly determined by the sealing materials and expressed by Coulomb friction as:

$$T_f = C_{cf} \cdot \mathrm{sgn}(\dot{\theta}) + C_{vf}\dot{\theta} \tag{9.3}$$

where C_{cf} and C_{vf} are the coefficients of Coulomb friction and viscous friction for rotary motion, respectively. $\mathrm{sgn}(\cdot)$ is a signum function.

In the case of a push motion, the flow mode is significantly developed under assumption of negligible shear mode effect. The produced force (F) can be described as summation of the controllable force F_c, viscous force F_η, restoring force F_r, and friction force F_f as:

$$F = F_c + F_\eta + F_r + F_f$$

$$= \frac{2d}{g}(A_d - A_r)\tau_y(\cdot) + \frac{12\eta h}{g^3 R}(A_d - A_r)^2\dot{x} + kx + F_f \tag{9.4}$$

where x and $\dot{x}$ are the displacement and velocity of the knob in push motion. A_d and A_r are the disk area and the rod area, respectively. k is the spring constant of the installed spring, which can restore the knob after pushing. F_f is the friction force by the uncontrollable sealing friction and expressed by

$$F_f = D_{cf} \cdot \mathrm{sgn}(\dot{x}) + D_{vf}\dot{x} \tag{9.5}$$

where D_{cf} and D_{vf} are the coefficients of Coulomb friction and viscous friction in push motion, respectively.

9.2.2 Design Optimization

In this section, finite element (FE) analysis is used to obtain the optimal geometric dimensions of the MR knob. Among several torque terms in Equation (9.2), the control torque has a large effect on the performance of the MR knob. It is desirable that the relative control torque to total torque takes a large value, which means the large dynamic control torque. Therefore, the objective function for optimal design is chosen to maximize the relative control torque, which is defined by

$$\lambda = \max\left\{\frac{T_c}{T_c + T_\eta}\right\} \tag{9.6}$$

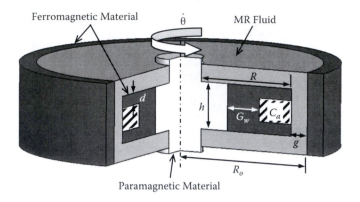

FIGURE 9.2
Geometric parameters of the MR knob.

It is noted that the optimality criterion does not include friction because it is not the function of geometry. The chosen design parameters to be optimized are the effective length of the disk except the magnetic coil (d), width of disk flux guide except the coil (G_w) and area of the coil (C_a). The significant geometric parameters are depicted on the cross-section view shown in Figure 9.2. In the figure, the light gray area is the MR fluid, the white area is the paramagnetic rod, the dark gray area is the ferromagnetic structures including electromagnetic disk and housing, and the slashed area is the coil.

The design parameters are optimally determined using ANSYS parametric design language to achieve the cost function, which is to maximize the relative control torque to the total torque including the control torque and fluid viscous torque. Because the knob geometry is axisymmetric, a 2D-axisymmetric coupled element (Plane 13) is used for the electromagnetic analysis. Moreover, the 4-node quadrilateral meshing is used for the FE model. The meshing size is specified by the number of elements per line rather than element size because the geometric dimensions of the knob vary during the optimization process.

For the optimization procedure, a log-file for solving the magnetic circuit of the MR knob and the relative control torque is built using the ANSYS parametric design language. In this log file, each design parameter is defined with its limits, tolerance, and initial value. After starting with the initial value of the design parameters, the average magnetic field inside the gap is calculated from the FE solution of the magnetic circuit. Once the yield stresses of the MR fluid inside the gap are obtained from the magnetic field, the objective function can be calculated by Equation (9.6). The ANSYS optimization tool then transforms the constrained optimization problem into the unconstrained one via penalty functions. The dimensionless unconstrained objective function is formulated as:

$$\lambda^*(x,q) = \frac{\lambda}{\lambda_0} + \sum_{i=1}^{n} P_x(x_i) + q \sum_{i=1}^{m} P_g(g_i) \tag{9.7}$$

where λ_0 is the reference objective function value that is selected from the current group of design sets and q is the response surface parameter, which controls constraint satisfaction. $P_x(x_i)$ is the exterior penalty function applied to the design variables x_i. $P_g(g_i)$ is the extended-interior penalty function applied to state variables g_i.

For the initial iteration ($i = 0$), the search direction is assumed to be the negative of the gradient of the unconstrained objective function. Thus, the direction vector is calculated by

$$D^{(0)} = -\nabla \lambda^*(x^{(0)}, 1) \tag{9.8}$$

where the superscript (0) stands for the number of iterations. The values of the design parameters in the ($i + 1$) iteration are obtained by

$$x^{(i+1)} = x^{(i)} + s_i D^{(i)} \tag{9.9}$$

where the line search parameter s_i is calculated using a combination of a golden-section algorithm and a local quadratic fitting technique. The log file is then executed with the new design values, and convergence of the objective function is checked. If convergence criteria is not satisfied, the subsequent iterations are performed with the direction vector calculated according to the Polak-Ribiere recursion formula [19] as:

$$D^{(j)} = -\nabla \lambda^*(x^{(j)}, q) + r_{j-1} D^{(j-1)}$$

$$r_{j-1} = \frac{\left| [\nabla \lambda^*(x^{(j)}, q) - \nabla \lambda^*(x^{(j-1)}, q)]^T [\nabla \lambda^*(x^{(j)}, q)] \right|}{\left| \nabla \lambda^*(x^{(j-1)}, q) \right|^2} \tag{9.10}$$

Thus, each iteration is composed of a number of sub-iterations that include search direction and gradient computations. The flowchart of the procedures to achieve optimal design parameters is shown in Figure 9.3.

Figure 9.4 shows the field-dependent yield stress of the employed MR fluid according to the flux density of the applied magnetic field. By applying a least-square curve-fitting method to the fluid properties, the approximated polynomial of the yield stress is given by

$$\tau_y(B) = 39.7 \times B^4 - 132.4 \times B^3 + 119.1 \times B^2 + 10.3 \times B + 0.1 \tag{9.11}$$

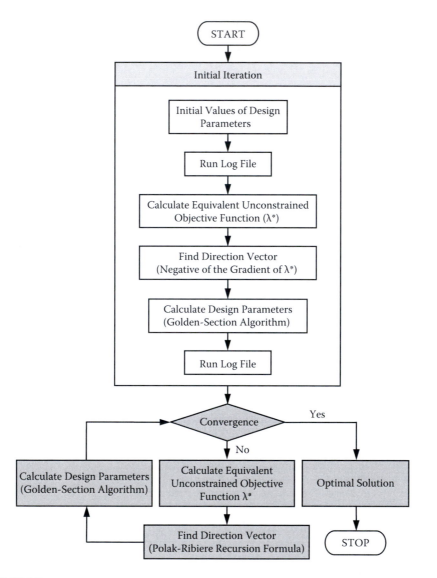

FIGURE 9.3
Flowchart of the optimal design procedure.

where B is the magnetic flux density. In addition, a mild steel (S45C) is adopted for the electromagnetic disk and housing as a ferromagnetic material. The rotational velocity of the disk is set at 140°/s. The outer radius and length of the electromagnetic disk is fixed at 30 mm and 15 mm, respectively, to consider the spatial limitation.

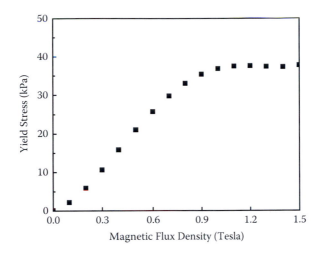

FIGURE 9.4
Properties of the employed MR fluid. (From Han, Y.M. et al., *Smart Materials and Structures*, 18, 1, 2009. With permission.)

Before starting the optimization procedure, the initial values for the design parameters of d, G_w, and C_a are assigned to be 2 mm, 9 mm, and 40 mm², respectively. Figure 9.5 shows the optimal solution of the MR knob in a specific spatial limitation when a current of 1.2 A is applied to the coil. From the figure, it is observed that the objective function converges after eight iterations and the corresponding torque is expected to be 0.704 Nm, which is 139% of the initial value. At the optimum, the design parameter values are shown in Figure 9.5(b), which are $d = 3.99$ mm, $G_w = 7.09$ mm, and $C_a = 56$ mm². The magnetic flux lines and the distribution of magnetic flux density along the gap are also shown in Figure 9.6. The average magnetic flux density by integrating along the defined path is 0.362 T. The design parameter values are finally determined to be $d = 4$ mm, $G_w = 7$ mm, and $C_a = 56$ mm² under consideration of manufacturing conditions. This makes the expected torque decrease to 0.692 Nm, but its difference from the optimal value is no more than 0.012 Nm. Figure 9.7 shows the exploded view and assembled view of the manufactured MR haptic device. A mild steel and duralumin are adopted as ferromagnetic and paramagnetic materials, and the number of coil turns corresponding to the optimized coil area ($C_a = 56$ mm²) is 100. Figure 9.8 presents its measured responses (scattered symbol) and expected responses (line) under rotary and push motions. From the results in Figure 9.8(a), the measured torque at input current of 1.2 A is 0.66 Nm, and its error to the expected torque of 0.692 is negligible. In addition, the measured force under push motion also agrees well with the expected ones as shown in Figure 9.8(b). The error between the measured force and the expected force at input current of 1.2 A is below 5.7 N.

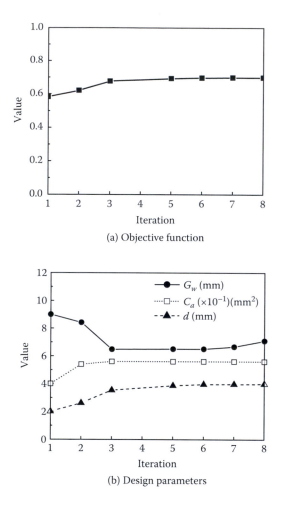

FIGURE 9.5
Optimization results of the MR haptic knob. (From Han, Y.M. et al., *Smart Materials and Structures*, 18, 1, 2009. With permission.)

9.2.3 Haptic Architecture

Figure 9.9 shows the haptic architecture that is composed of the manufactured MR haptic knob, virtual in-vehicle functions, torque/force reflection algorithm, and driver. The manufactured MR haptic knob interacts with a virtual environment of vehicle through position and force, which come from the real MR knob via operation and the virtual environment via a force reflection algorithm. In the meantime, a driver operating the MR knob undergoes execution states of the corresponding in-vehicle function from the virtual environment. In other words, when a person operates the MR knob, the measured position information is transferred to the virtual environment. Then the corresponding torque/force map in

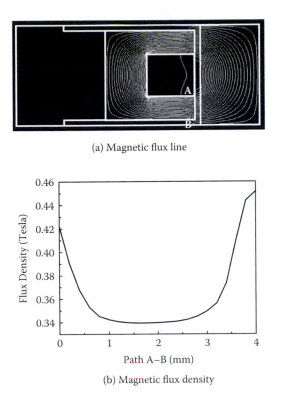

(a) Magnetic flux line

(b) Magnetic flux density

FIGURE 9.6
FE results of the optimized MR knob. (From Han, Y.M. et al., *Smart Materials and Structures*, 18, 1, 2009. With permission.)

the force reflection algorithm generates the desired torque or force for the MR knob to track.

The virtual environment is established by LabVIEW under consideration of various in-vehicle functions within a real passenger vehicle. Figure 9.10 shows the virtual environment, which has the data display windows for input and output measurements, and four in-vehicle function menus: heater/air conditioner, entertainment, window, and seat. Each function has several sub-events, which can be expressed by a tree menu with three levels as shown in Figure 9.11.

After selecting the event in the virtual environment, the corresponding torque/force trajectories are determined by eliminating the fluid viscous term and the friction term from the torque/force map upon the measured position information as:

$$\tilde{T}_d = \tilde{T}(\theta) - (4\pi\eta R^3 h/g)\dot{\theta} - T_f(\dot{\theta})$$

$$\tilde{F}_d = \tilde{F}(x)g - \frac{12\eta h}{g^3 b}(A_p - A_r)^2 \dot{x} - F_f(\dot{x})g - kx \qquad (9.12)$$

Housing   Operational
                            Knob

Electromagnetic Disk

(a) Exploded

(b) Assembled

FIGURE 9.7
The manufactured MR knob.

where $\tilde{T}(\theta)$ and $\tilde{F}(x)$ are the calculated trajectories from the torque/force map upon optional input. In order to achieve the desired trajectories in Equation (9.12), a feed-forward controller can be formulated by inversion of the following controllable torque and force models.

$$F_c = \frac{2d}{g}(A_d - A_r) \cdot \alpha H_F^\beta$$

$$T_c = 4\pi R^2 d \cdot \alpha H_T^\beta$$

(9.13)

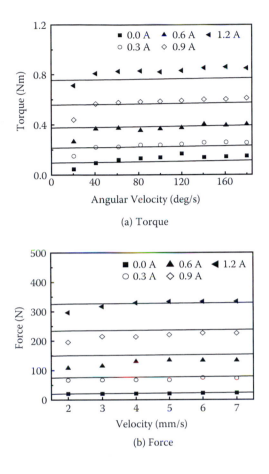

FIGURE 9.8
Measured responses of the manufactured MR knob. (From Han, Y.M. et al., *Smart Materials and Structures*, 18, 1, 2009. With permission.)

Therefore, the control input can be expressed in terms of magnetic field strength as:

$$H_T = \left(\frac{\tilde{T}_d}{\alpha \cdot 4\pi R^2 d} \right)^{\frac{1}{\beta}}$$

$$H_F = \left(\frac{\tilde{F}_d}{\alpha \cdot 2d(A_p - A_r)} \right)^{\frac{1}{\beta}} \qquad (9.14)$$

where α and β are the intrinsic values of the MR fluid, which are determined by the experimental Bingham model of $0.092H^{1.236}$. H_T and H_F are the

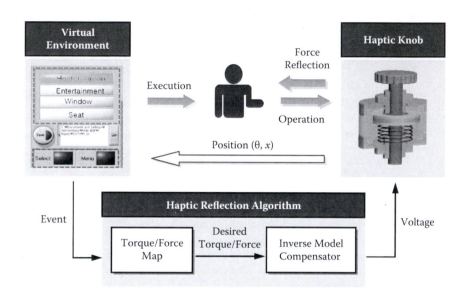

FIGURE 9.9
Haptic architecture.

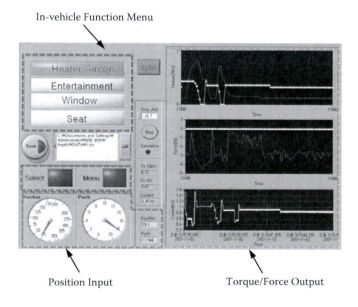

FIGURE 9.10
Virtual environment.

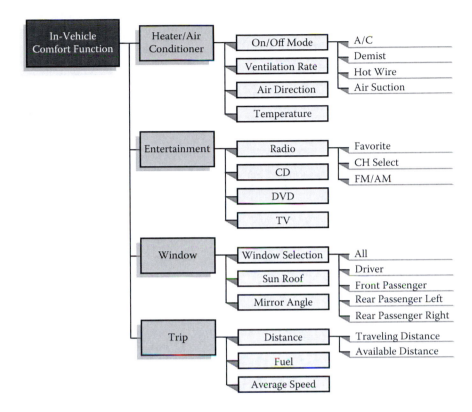

FIGURE 9.11
Tree menu of the virtual in-vehicle functions.

input magnetic field strengths for rotary motion and push motion, respectively, as control inputs. The magnetic field strength can be determined by $NI/2g$ where I is electric current, N is the number of coil turns, and g is the gap between the electromagnetic disk and housing. Once the control input is determined, the input current to be applied to the MR knob is obtained as

$$I_T = \frac{2g}{N} \left(\frac{\tilde{T}_d}{\alpha \cdot 4\pi R^2 d} \right)^{\frac{1}{\beta}}$$

$$I_F = \frac{2g}{N} \left(\frac{\tilde{F}_d}{\alpha \cdot 2d(A_p - A_r)} \right)^{\frac{1}{\beta}}$$

(9.15)

Figure 9.12 shows the corresponding control block diagram. It is noted that the MR fluid is modeled by the simple Bingham model. This is reasonable because the torque and force level is small, and the control input level is also small. In general, the high current level causes a large hysteresis loop for the force-velocity relationship. However, the input current level of the multi-functional

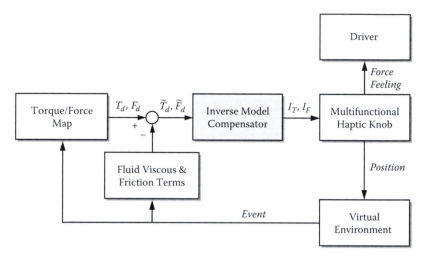

FIGURE 9.12
Block diagram of the feed-forward controller.

MR control knob is not as high as the corresponding force-velocity hysteretic behavior, which seriously affects modeling accuracy at the operating velocity of vehicular instrument controls around 100°/s. Therefore, complicated models of MR fluid such as the Preisach model or the biviscous model are not required to capture the hysteretic behavior of the MR fluid. Furthermore, the proposed simple controller using the Bingham model can be an effective candidate for practical implementation to vehicle engine control unit (ECU).

9.2.4 Performance Evaluation

Figure 9.13 shows the experimental configuration to evaluate control performance of the proposed multi-functional MR haptic knob that interacts with

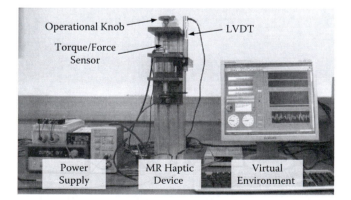

FIGURE 9.13
Experimental apparatus.

the virtual environment realized by LabVIEW. The knob connected to the electromagnetic disk is operated by a person who corresponds to a vehicle driver. When he or she operates the MR knob, the position information is obtained from the encoder and linear variable differential transducer (LVDT). A specific in-vehicle function in the virtual environment is then visually operated and generates the desired torque or force upon the position information by the map. The feed-forward controller given by Equation (9.13) is then activated for the MR knob to reflect the desired torque or force to a driver. The produced torque and force are measured by a 6-axis force/torque transducer (ATI, Nano25).

Among several events of the in-vehicle functions, three events of menu shifting, temperature adjusting, and window opening are considered for demonstration of the haptic control performance of the MR knob. Figure 9.14 shows the torque/force map adopted for the multi-functional MR control knob. The corresponding torque/force profile for each event is defined according to the positional inputs. Figure 9.14(a) shows the torque map adopted for menu shifting. The event of menu shifting occurs at transition from one menu to the next or the previous menu. Therefore, a click torque should be produced at specific angles for a driver to recognize the transition between two menus by tactile feeling. The event of menu shifting occurs at 60°, 120°, and 180°, which are the shifting positions from heater/air conditioner to entertainment, from entertainment to window, and from window to seat, respectively. The whole operating range is from 0° to 240° and the stopping torque of 0.8 Nm is applied at both ends of the operating range. In addition, the click torque is required for both clockwise (CW) and counterclockwise (CCW) directions of rotation at the same position. Figure 9.14(b) shows the torque map for the event to adjust temperature of a vehicle cabin. Several triangular-like profiles of click torque are defined at every 10 degrees between 0° and 200°. Peak magnitude of each triangle increases from 0 Nm to 0.8 Nm by 0.08 Nm as rotational angle so that a driver can recognize temperature change from 20°C to 30°C by tactile feeling without visual attention. Figure. 9.14(c) shows the force map for the event of window opening in push motion. Its profile linearly increases from 0 N to 80 N for the pushed displacement of 20 mm in a forward direction. In this case, according to the displacement, the window opening position is adjusted between 0 cm and 40 cm, whose corresponding force between 0 N and 80 N is reflected to a driver.

Figure 9.15 shows the torque tracking control results for the menu shifting, which is an event of first-level function in the menu tree. As clearly observed from the results in Figure 9.15(b), the desired torque from the virtual environment is well generated by the positional input shown in Figure 9.15(a) and favorably followed by the feed-forward controller with tracking errors below 0.09 Nm, which means that the relative error is 16.4%. In addition, sufficient stopping torque up to 0.73 Nm is achieved at 0° and 240°. Figure 9.16 shows results for the event of temperature

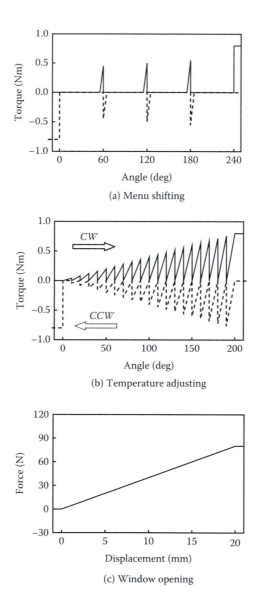

FIGURE 9.14
Torque/force map. Torque/force map. (From Han, Y.M. et al., *Smart Materials and Structures*, 18, 1, 2009. With permission.)

adjusting in rotary motion, which is an event of second-level function in the heater/air conditioner category. According to the angular displacement of the knob in Figure 9.16(a), the torque trajectory of click is generated as a desired trajectory. Figure 9.16(b) presents that the MR knob well reflects the feeling of clicks for both CW and CCW directions with

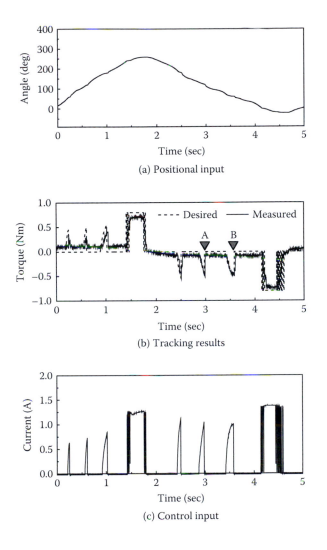

FIGURE 9.15
Control results for the event of menu shifting. (From Han, Y.M. et al., *Smart Materials and Structures*, 18, 1, 2009. With permission.)

appropriate input current histories in Figure 9.16(c). It is noted that a time delay exists in Figure 9.16 (b). In particular, this phenomenon becomes worse while the rotational direction is reversed. The change of rotation direction may cause vortex of the ER fluid in an electrode gap. But the click torque in one rotational direction (CW or CCW) shows a very small time delay below 30 ms. Figure 9.17 shows force tracking control results for the event of the driver's window opening in push motion, which is an event of third-level function in the window category. From the results shown in Figure 9.17(b), it is observed that the desired force is well generated by the

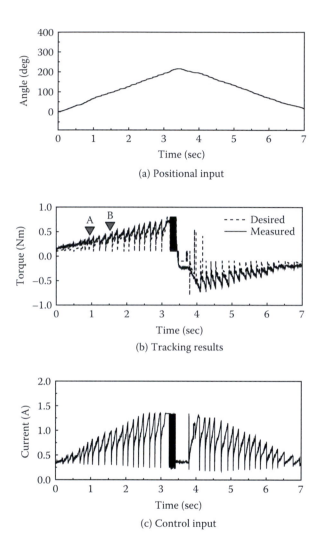

(a) Positional input

(b) Tracking results

(c) Control input

FIGURE 9.16
Control results for the event of temperature adjusting. (From Han, Y.M. et al., *Smart Materials and Structures*, 18, 1, 2009. With permission.)

virtual in-vehicle function and achieved from the MR knob by activating the controller. The calculated tracking errors are below 7.9 N after 1s, and the corresponding relative error is 18.2%. Figure 9.17(c) shows the corresponding input histories of current applied to the MR knob.

In addition, operations of the virtual environment are graphically visualized in Figure 9.18. Figure 9.18(a) shows resultant virtual images of menu shifting. The corresponding positions are marked as "A" and "B" in Figure 9.15(b). As recognized from the maximum reflected torque

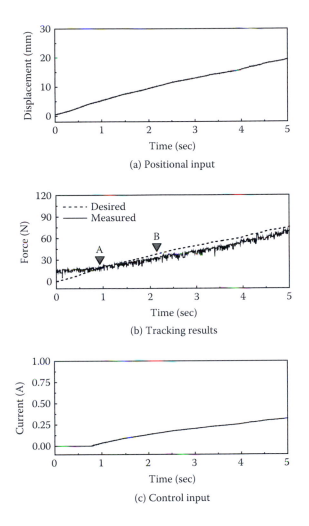

FIGURE 9.17
Control results for the event of window opening. (From Han, Y.M. et al., *Smart Materials and Structures*, 18, 1, 2009. With permission.)

of −0.46 Nm, the menu is shifted from entertainment to heater/air conditioner. Figure 9.18(b) shows resultant images of a virtual environment for temperature adjusting at two distinct position marked as "A" and "B" in Figure 9.16(b). From the results, it is observed that the temperature is well adjusted to 24°C and 26°C while a person feels the click force of 0.40 Nm and 0.54 Nm at the knob angle of 80° and 120°, respectively. Figure 9.18(c) shows virtual images for the window opening. At two marked positions of "A" and "B" in Figure 9.17(b), the driver's window is opened by 12 cm and 20 cm, respectively, while the corresponding force of 17.7 N and 28.9 N are reflected to the driver.

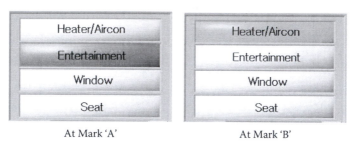

(a) Menu shifting

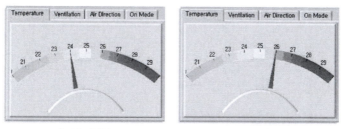

(b) Temperature adjusting

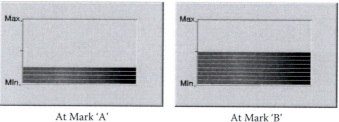

(c) Window opening

FIGURE 9.18
Virtual operation of the in-virtual function event.

9.3 MR Haptic Cue Accelerator

9.3.1 Configuration and Optimization

When driving automatic transmission vehicles, the vehicle automatically starts in low gear, and then goes up to top gear. But in manual transmission, the driver should decide the proper moment of gear shifting, that which is significantly related to driving power and fuel consumption. In order to transmit the optimal moment of gear shifting to a driver, accelerator pedal is considered because the most meaningful gear shifting occurs when

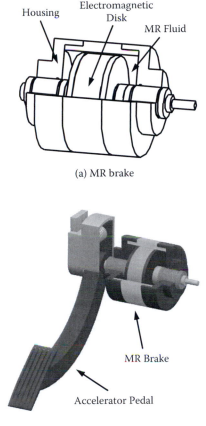

(a) MR brake

(b) Haptic cue accelerator

FIGURE 9.19
Schematic configuration of the MR haptic cue accelerator.

accelerating the vehicles. Figure 9.19 shows the schematic configuration of the proposed MR haptic cue device with accelerator pedal, which is devised by adopting a rotary type of brake mechanism. The accelerator pedal is connected to a cylindrical electromagnetic disk, which has a flux guide and an electromagnetic coil. The electromagnetic disk is assembled to housing with a specific gap fully filled with MR fluid. The MR product (MRF-132DG) [18] from the Lord Corporation has been employed as a controllable fluid. Again remembering the Bingham model of the MR fluid in Chapter 1, the shear stress (τ) under a magnetic field is expressed by

$$\tau = \tau_y(\cdot) + \eta\dot{\gamma} \qquad (9.16)$$

where η is the dynamic viscosity, γ is the shear rate, and $\tau_y(\cdot)$ is the dynamic yield stress of the MR fluid. It is noted that the applied magnetic field can be expressed by one of magnetic flux density (*B*) and magnetic field strength (*H*), and they are used for design optimization and controller synthesis, respectively.

Among several operation modes of MR fluid-based devices, the shear mode is significantly developed for the rotary motion. The produced torque is mathematically described as a summation of the following controllable torque T_c, viscous torque T_η, and friction torque T_f:

$$T = T_c + T_\eta + T_f$$

$$= 4\pi R^2 d\tau_y(\cdot) + \frac{4\pi\eta R^3 h\dot{\theta}}{g} + T_f \qquad (9.17)$$

where $\dot{\theta}$ is the rotational velocity of the haptic cue device in rotary motion. *R* and *h* are the outer radius and length of the electromagnetic disk, respectively. *d* is the length of the electromagnetic disk except for the width of the magnetic coil. *g* is the gap between the disk and housing. As shown in the previous equation, the produced torque (*T*) has an additional term, friction torque T_f. The friction torque is mainly determined by the uncontrollable sealing friction and expressed by Coulomb friction as:

$$T_f = C_{cf} \cdot \text{sgn}(\dot{\theta}) + C_{vf}\dot{\theta} \qquad (9.18)$$

where C_{cf} and C_{vf} are the coefficients of Coulomb friction and viscous friction for rotary motion, respectively. sgn($\cdot$) is a signum function.

Figure 9.20 shows the cross-section view of the haptic cue device. In the figure, the light gray area is the MR fluid, the white area is the paramagnetic

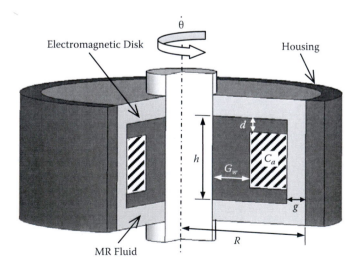

FIGURE 9.20
Geometric parameters of the MR haptic cue device.

rod, the dark gray area is the ferromagnetic structures including electromagnetic disk and housing, and the slashed area is the coil. The electromagnetic disk rotates inside of the housing according to the pushed rotation of the accelerator pedal. Then, the applied magnetic field by the coil produces a resistance torque to the pedal motion. It is noted that this resistance torque is just used to transmit a cue signal to the driver. So its magnitude is much smaller than the level of the driver's pushing torque and restoring torque of the accelerator pedal so as not to go against the driver's intension.

In this section, the geometric dimensions of the MR haptic cue accelerator is optimally determined using the ANSYS parametric design approach. The objective function for the haptic cue accelerator is chosen in the same manner as the multi-functional knob in Equation (9.6), which is to maximize the relative control torque to the total torque including the control torque and fluid viscous torque. The chosen design parameters to be optimized are the half length of the disk except the magnetic coil (d), width of disk flux guide except the coil (G_w) and the area of the coil (C_a).

The optimization procedure follows the flowchart in Figure 9.3. After starting with the initial value of the design parameters, the equivalent unconstrained objective function in Equation (9.7) is recursively calculated until the design parameters converge to optimum. This recursive procedure includes calculation of the average magnetic field inside the gap from the FE solution of the magnetic circuit and the resultant yield stress caused by the MR fluid. Figure 9.4 shows the experimental field-dependent yield stress of the employed MR fluid according to the flux density of the applied magnetic field. The outer radius and length of the electromagnetic disk is fixed to be 40 mm and 35 mm, respectively, to consider the spatial limitation.

Before starting the optimization procedure, the initial values for the design parameters of d, G_w, and C_a are assigned to be 4 mm, 12 mm, and 378 mm^2, respectively. Figure 9.21 shows the optimal solution in a specific spatial limitation when a current of 0.6 A is applied to the coil. From Figure 9.21(a), it is observed that the objective function converges after seven iterations and the corresponding torque is expected to be 6.47 Nm, which is about 160% of the initial value. During optimization, the varying design parameter values are show in Figure 9.21(b), and the optimum values are $d = 6.49$ mm, $G_w = 11.21$ mm, and $C_a = 325$ mm^2. The magnetic flux lines and the distribution of magnetic flux density along the gap are also shown in Figure 9.22. The maximum flux density is 1.825 T. The values of design parameters are finally determined to be $d = 6$ mm, $G_w = 11$ mm, and $C_a = 345$ mm^2 under consideration of manufacturing convenience.

Figure 9.23(a) shows the exploded view of the manufactured MR haptic cue device, which is composed of the housing and the electromagnetic disk. A mild steel (S45C) is adopted for the flux guide as ferromagnetic material. The other part was made of paramagnetic aluminum (Al6061). The optimized coil area

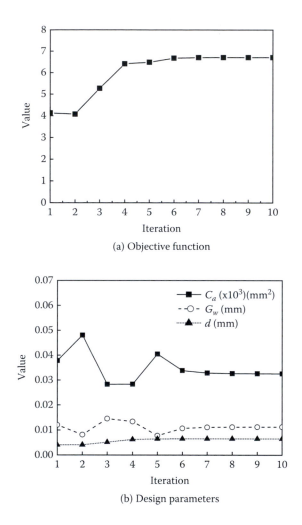

FIGURE 9.21
Optimization results of the MR haptic cue device. (From Han, Y.M. et al., *Smart Materials and Structures*, 19, 7, 2010. With permission.)

(C_a = 345 mm²) was winded by copper wire with 0.75 mm diameter. Its number of coil turns is 500. The electromagnetic disk was assembled to the housing with a specific gap fully filled with the MR fluid (MRF-132DG) from the Lord Corporation. Figure 9.23(b) shows the assembled view of the manufactured MR haptic cue device, which is connected with a commercial accelerator pedal. Table 9.1 presents design parameters of the manufactured MR haptic cue device.

Figure 9.24 presents the produced torque characteristics of the manufactured haptic cue device. A variable speed AC motor with controller is used to drive. A rotational torque transducer, which has a capacity of 10 Nm, is employed to measure the produced torque. An angular position is

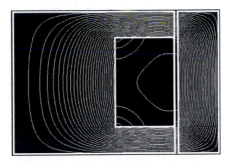

(a) Magnetic flux line

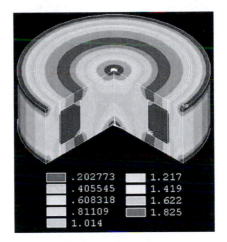

.202773	1.217
.405545	1.419
.608318	1.622
.81109	1.825
1.014	

(b) Magnetic flux density

FIGURE 9.22
FE analysis of the optimized MR haptic cue device. (From Han, Y.M. et al., *Smart Materials and Structures*, 19, 7, 2010. With permission.)

measured by an incremental encoder, which has resolution of 3600 pulses per revolution. Figure 9.24(a) shows the field-dependent torque according to eight different angular velocities between 10°/s and 80°/s. The measured responses (scatter symbol) are compared with the predicted responses (line). Under input current of 0.6 A and rotational velocity of 10°/s, the MR haptic cue device produces 6.67 Nm. It is well matched with the optimal result from FE analysis. Figure 9.24(b) shows the total reaction torque of the accelerator pedal connected with the MR haptic device. The total torque is a little higher than the torque from Figure 9.24(a) because a restoring spring inside the pedal produces additional torque. From the results, it can be assured that the reaction torque of the accelerator pedal can be continuously controlled by adopting the proposed MR haptic device and appropriate control scheme.

(a) Exploded

(b) Assembled (with accelerator pedal)

FIGURE 9.23
The manufactured MR haptic cue accelerator.

9.3.2 Automotive Engine-Transmission Model

Figure 9.25 shows the engine-transmission model. A triggered subsystem considering manifold dynamics transfers the air-fuel mixture from the intake manifold to the cylinders via discrete valve events. Then a four-cylinder spark ignition internal combustion engine generates torque and accelerates the vehicle body through the transmission system by

TABLE 9.1

Design Parameters of the MR
Haptic Cue Device

Design Parameter	Value
Height of flux guide (d)	0.006 m
Area of coil (C_a)	345 mm²
Width of flux guide (G_w)	0.011 m
Gap size (g_d)	0.001 m
Radius of disk (R)	0.04 m
Coil turns (N)	500

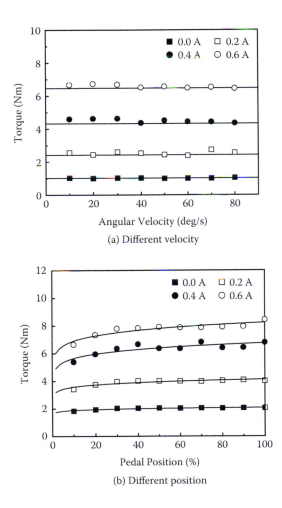

(a) Different velocity

(b) Different position

FIGURE 9.24

Toque characteristics of the manufactured MR haptic cue device. (From Han, Y.M. et al., *Smart Materials and Structures*, 19, 7, 2010. With permission.)

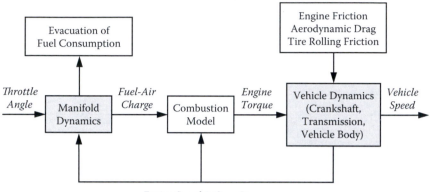

FIGURE 9.25
Engine-transmission model.

vehicle dynamics. The continuous-time processes of intake flow take place concurrently with the torque generation and vehicle acceleration. The adopted engine-transmission model considers two main subsystems such as throttle-manifold dynamics and vehicle dynamics. There exists a throttle plate inside the throttle body, and its angle is the input variable. The throttle angle increases to supply air including fuel to the engine when a driver pushes the accelerator pedal. The air-fuel charge rate through the throttle body is a function of the throttle angle ϕ (degree) and manifold pressure p (kPa). Therefore, the mass flow rate into manifold $\dot{m}$ is given by [20, 21]

$$\dot{m} = f(\theta) \cdot g(p) \tag{9.19}$$

where

$$f(\theta) = 2.821 - 0.05231\,\theta + 0.10299\,\theta^2 - 0.00063\,\theta^3$$

$$g(p) = \begin{cases} 1, & \text{if } p < 0.5 p_{atm} \\ \dfrac{2}{p_{atm}}\sqrt{p_{atm}p - p^2}, & \text{if } p \geq 0.5 p_{atm} \end{cases}$$

In the above, p_{atm} is the atmospheric pressure, which is approximately 100 kPa.

The difference in the incoming and outgoing mass flow rates represents the net rate of change of air mass with respect to time. This quantity, according to the ideal gas law, is proportional to the time derivative of the manifold pressure. Therefore, the manifold dynamics are usually modeled as a first-order linear system as:

$$\dot{p} = k(\dot{m} - \dot{M}) \tag{9.20}$$

where $k = 0.5786$, which is determined from specific gas constant, temperature, and manifold volume under idealized assumptions. $\dot{M}$ is the mass flow rate into the cylinder given by [20, 21]

$$\dot{M} = -0.366 + 8.979979pn - 337np^2 + 0.01n^2p \tag{9.21}$$

where n is the angular velocity of the engine. The air-fuel mixture is charged into the cylinder during the intake stroke, which takes place in the first π rotation of the crankshaft in the four-stroke cycle: the intake, compression, combustion, and exhaust strokes. This air-fuel mixture charged into the cylinder can be used to evaluate the fuel consumption of the vehicles adopting the proposed haptic cue accelerator.

For a four-cylinder four-stroke engine, the ignition of each successive cylinder is separated by π. The total air-fuel mass pumped into the cylinders can be determined by integration of the mass flow rate from the intake manifold at the end of each intake stroke event before compression. Thus, the air-fuel charge for each individual cylinder per intake stroke is given by

$$M_c = \dot{M} \frac{\pi}{n} \tag{9.22}$$

Now the torque generated by the engine for combustion stroke can be described by the following empirical function [20, 21]:

$$T_{eng} = -181.3 + 379.36M_c + 21.91(A/F) - 0.85(A/F)^2 + 0.26\sigma - 0.0028\sigma^2$$
$$+ 0.027n - 0.000107n^2 + 0.00048n\sigma + 2.55\sigma M_c - 0.05\sigma^2 M_c \tag{9.23}$$

where (A/F) is the air-to-fuel mixture ratio, which in our case is 14.6. σ is the spark advance, which can be selected to maximize the engine torque as:

$$\sigma = \frac{0.26 + 0.00048n + 2.55M_c}{0.0056 + 0.1M_c} \tag{9.24}$$

Under the external force F, the dynamic behavior of the vehicle body can be simply expressed by

$$M_v \dot{v} = F \tag{9.25}$$

where $\dot{v}$ is the acceleration of the vehicle body and M_v is the mass of the vehicle body. However, the engine power is first transferred to rotate the engine crankshaft, which is connected with the vehicle wheels. Therefore, by considering the rotational inertia effect of the crankshaft, Equation (9.25) can be changed as:

$$J\dot{n} + M_v\dot{v} = F \tag{9.26}$$

where $\dot{n}$ is the acceleration of the engine, and J is the rotational moment of inertia of the engine.

The engine power is transferred to the vehicle body through the transmission system, which should be considered in the model. Therefore, the vehicle speed is related to the engine speed as:

$$v = r_w g_r(i)n, \quad i = 1,2,3,4 \tag{9.27}$$

where r_w is the radius of the vehicle wheels, and $g_r(i)$ is the transmission gear ratio for ith gear position. In this case, the transmission is assumed to have five gear positions with the corresponding ratios shown in Table 9.2.

In addition, the external force to drive the vehicle can be expressed with the engine torque in Equation (9.23) as:

$$F = \frac{1}{r_w g_r(i)}[T_{eng} - T_l] \tag{9.28}$$

where T_l is the load torque, which can be expressed as a function of the drag due to the engine friction, tire rolling friction, and aerodynamic drag as:

$$T_l = C_{ce} \cdot \text{sgn}(n) + C_{ve}n + \mu_t M_v g r_w g_r(i) + 0.5 C_d \rho A_v [r_w g_r(i)]^3 n^2 \tag{9.29}$$

where C_{ce} and C_{ve} are the coefficients of Coulomb friction and viscous friction for rotary motion of the vehicle engine, respectively. μ_t is the coefficient of kinetic friction of the vehicle tire, and g is the acceleration of gravity. C_d is the coefficient of aerodynamic drag, ρ is the air density, and A_v is the

TABLE 9.2

Transmission Gear Ratio

Gear Stage	Gear Ratio	Differential Gear Ratio	Reduction Ratio
1st Gear	3.818		0.0589
2nd Gear	2.210		0.1018
3rd Gear	1.423	4.444	0.1581
4th Gear	1.029		0.2187
5th Gear	0.837		0.2688

TABLE 9.3

Mechanical Parameters for the Adopted Vehicle

Parameter	Value
Vehicle mass (M_v)	830 kg
Tire radius (r_w)	260 mm
Aerodynamic drag coefficient (C_d)	0.34
Frontal cross-sectional area (A_v)	2.02 m²
Air density (ρ)	1.29 kg/m³

frontal cross-sectional area of the vehicle. In this test, a small-sized vehicle is adopted, and its mechanical parameters are shown in Table 9.3.

Consequently, the vehicle dynamics including the vehicle body, transmission, and crankshaft can be rewritten as:

$$\left\{ J + \frac{M_v}{[r_w g_r(i)]^2} \right\} \dot{n}$$

$$= \frac{1}{r_w g_r(i)} \left\{ \begin{array}{l} -181.3 + 379.36 M_c + 21.91(A/F) - 0.85(A/F)^2 + 0.26\sigma - 0.0028\sigma^2 \\ + 0.027n - 0.000107n^2 + 0.00048n\sigma + 2.55\sigma M_c - 0.05\sigma^2 M_c \end{array} \right\}$$

$$- \frac{1}{r_w g_r(i)} \left\{ C_{ce} \cdot \mathrm{sgn}(n) + C_{ce} n + \mu_t M_v g r_w g_r(i) + 0.5 C_d \rho A_v [r_w g_r(i)]^3 n^2 \right\}$$

(9.30)

9.3.3 Haptic Architecture

Based on the engine-transmission model, a virtual vehicle is established to determine a gear shifting timing and evaluate fuel consumption of vehicles adopting the proposed MR haptic cue system. Figure 9.26 shows the virtual environment of the vehicle established by MATLAB Simulink® modeling techniques. The virtual vehicle emulates a four-cylinder four-stroke engine and manual transmission system of a passenger vehicle to determine the gear shifting timing based on the engine speed. Its environment is composed of engine speed, pedal angle, gear stage, and data display windows for other inputs and outputs.

Figure 9.27 shows the haptic architecture composed of the MR haptic cue accelerator and virtual environment of a passenger vehicle. The MR haptic cue accelerator interacts with the virtual vehicle through driver and control algorithm. When a driver pushes the accelerator pedal, its positional information is transmitted to the virtual vehicle, which generates cue signals for gear shifting. The input current is then determined with the torque control algorithm and cue signal. Finally, the MR haptic cue accelerator reflects reaction

Engine Speed   Pedal Angle   Gear Position

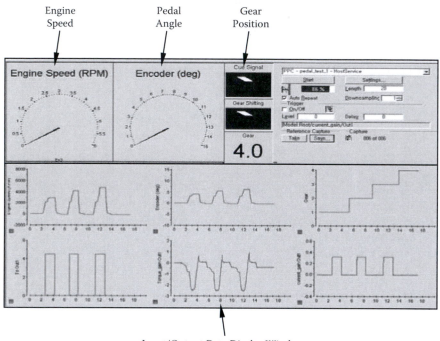

Input/Output Data Display Window

FIGURE 9.26
Virtual environment.

torque to the driver in order to notify appropriate gear shifting timing. In the meantime, the driver can determine whether to follow the suggested timing.

The cue signal for optimal gear shifting is generated at a specific engine speed. After generating a cue signal in the virtual environment, the desired reaction torque is determined from a torque map, which has the relationship between the reaction torque and the engine speed. However, the torque model in Equation (9.17) has the uncontrollable terms such as the viscous term and the friction term. Therefore, the controllable desired torque trajectory T_{cd} is determined by eliminating the fluid viscous torque term, friction torque term, and pedal torque term from the torque map output upon the measured position information as:

$$T_{cd} = T_d - \frac{4\pi\eta R^3 h}{g}\dot{\theta} - C_{cf}\cdot\mathrm{sgn}(\dot{\theta}) + C_{vf}\dot{\theta} - T_r(\theta) \qquad (9.31)$$

where T_d is the desired reaction torque from the torque map. $T_r(\theta)$ is the restoring torque by the spring element of the pedal, which is given by:

$$T_r(\theta) = T_i + K_r\alpha_r\, l \qquad (9.32)$$

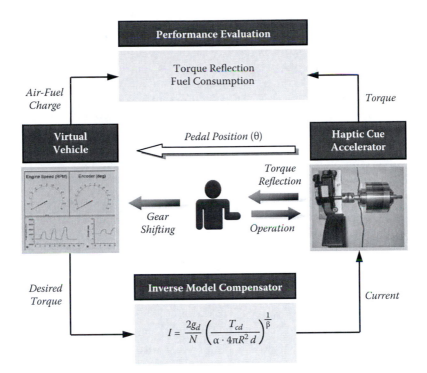

FIGURE 9.27
Architecture of the proposed haptic cue system.

where K_r and α_r are the spring constant and angular displacement of the accelerator pedal, respectively. T_i is the initial restoring torque of the pedal. l is the effective length of the pedal arm.

A feed-forward controller can be formulated by inversion of the controllable torque model. From Equation (9.17), the controllable torque T_c is given by

$$T_c = 4\pi R^2 d\tau_y(H) \tag{9.33}$$

In the above, the dynamic yield stress of the MR fluid can be expressed by magnetic field strength (H) as:

$$\tau_y(H) = \alpha H^\beta \tag{9.34}$$

where α and β are the intrinsic values of the MR fluid, which are determined by the experimental Bingham model: $0.092H^{1.236}$.

In order to achieve the controllable desired torque trajectory, the controllable torque T_c should follow T_{cd} as:

$$T_{cd} = 4\pi R^2 d \cdot \alpha H^\beta \tag{9.35}$$

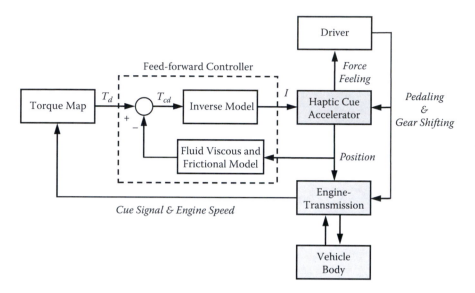

FIGURE 9.28
Block diagram of the feed-forward controller.

By inversion of the above model, the control input can be expressed in terms of magnetic field strength as:

$$H = \left(\frac{T_{cd}}{\alpha \cdot 4\pi R^2 d} \right)^{\frac{1}{\beta}} \tag{9.36}$$

Once the magnetic field strength as control input is determined, the input current to be applied to the MR haptic cue device can be calculated by [22]:

$$I = \frac{2g}{N} \left(\frac{T_d - \dfrac{4\pi\eta R^3 h}{g}\dot{\theta} - C_{cf} \cdot \mathrm{sgn}(\dot{\theta}) + C_{vf}\dot{\theta} - F_i - K_r\alpha_r\, l}{\alpha \cdot 4\pi R^2 d} \right)^{\frac{1}{\beta}} \tag{9.37}$$

where N is the number of coil turns. Figure 9.28 shows the corresponding control block diagram.

9.3.4 Performance Evaluation

Figure 9.29 shows the experimental apparatus to evaluate control performance of the proposed MR accelerator that interacts with the virtual vehicle realized by dSPACE and MATLAB Simlink®. The accelerator pedal connected to the electromagnetic disk of the MR device is pushed by a cam system driven by an AC motor, which corresponds to a vehicle driver. When the

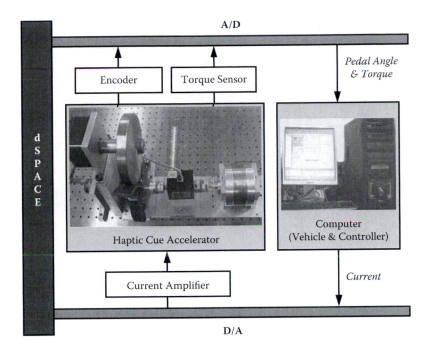

FIGURE 9.29
Experimental apparatus for the haptic cue control.

accelerator pedal is pushed, the pedal position is obtained by an incremental encoder, which has a resolution of 3600 pulses per revolution. The throttle angle increases to supply air including fuel into the engine, and its mapping ratio to pedal angle is 3.0. A four-cylinder four-stroke engine is then virtually operated and generates the driving torque, which is transferred to the vehicle body through the transmission system. In the meantime, a cue signal for gear shifting is generated according to the engine speed. In this test, the cue signal is set to occur at 2500 rpm of engine speed. Then the desired reaction torque is determined from the torque model and torque map. The feed-forward controller in Equation (9.37) is then activated for the MR haptic cue device to reflect the desired torque to a driver. The driver is notified of the optimal moment of gear shifting and the virtual environment shows the execution of gear shifting. In the meantime, the consumed fuel was evaluated with the manifold dynamics in Equation (9.20).

Figure 9.30 shows the employed torque map when accelerating a vehicle from low gear to high gear. The output torque of the map was determined based on the engine speed. Its threshold is 2500 rpm at which the gear is expected to change. After a cue signal for the gear shifting at 2500 rpm of engine speed, a normal reaction torque (below 2 Nm) changed to a constant torque of 4.5 Nm in order to cue the driver to change into high gear. It is noted that the reaction torque is only kept for a few seconds whether a driver changes the gear or not.

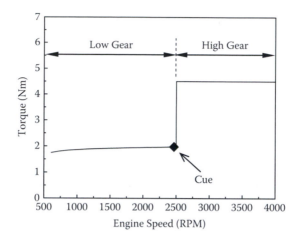

FIGURE 9.30
Torque map for the gear shifting. (From Han, Y.M. et al., *Smart Materials and Structures*, 19, 7, 2010. With permission.)

Figure 9.31 presents torque tracking results for gear shifting from second gear to third gear. In the beginning of the second gear stage, the vehicle cruises in a steady speed at which engine and vehicle speeds are approximately 1800 rpm and 5 m/s, respectively. It is assumed that it takes approximately 0.5 s until the driver recognizes the haptic cue and tries to change the gear stage. Therefore, the accelerator pedal was released after 0.5 s from the cue signal. By pushing the pedal to change the gear, throttle angle increased and the rotational speed of the vehicle engine was determined as shown in Figure 9.31(a) from the engine model in Equation (9.30). When the engine speed meets the threshold of 2500 rpm, a cue signal and the corresponding desired torque trajectory are generated. Then the MR haptic cue device is activated to follow them. As clearly observed from the results in Figure 9.31(b), the desired torque trajectory from the map is well followed by the feed-forward controller after cue signal. The response time of the proposed device is favorable and the final tracking error before releasing the pedal is below 0.21 Nm. In the meantime, input current is shown in Figure 9.31(c) whose magnitude is approximately 0.32 A. It is noted that the initial torque about 1 N/m at 2 s corresponds to the normal reaction torque of the accelerator pedal by friction and spring. The engine speed exceeds 2500 rpm a little bit because of 0.5-s delay for gear change as mentioned before.

In order to show the feasible operation of the proposed haptic cue accelerator, a driver-based haptic cue was performed, as shown in Figure 9.32. During an experiment, a voluntary human operator pushed the pedal and felt the reaction torque. At the same time, the reflected torque at the pedal was measured by the torque sensor. Based on the reflected torque through the pedal, the driver changed the gear from first to fourth in sequential manner. As

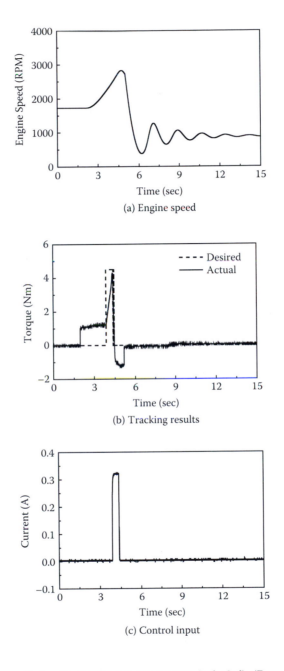

FIGURE 9.31
Torque tracking results for the haptic cue (gear stage: 2nd→3rd). (From Han, Y.M. et al., *Proceedings of the Institution of Mechanical Engineers : Part D - Journal of Automobile Engineering,* 225, 3, 2011. With permission.)

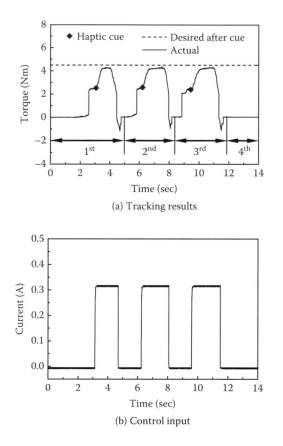

(a) Tracking results

(b) Control input

FIGURE 9.32
Driver-based haptic cue from 1st gear to 4th gear. (From Han, Y.M. et al., *Proceedings of the Institution of Mechanical Engineers : Part D - Journal of Automobile Engineering*, 225, 3, 2011. With permission.)

clearly observed from the control results, the torque was successfully reflected to the human driver by activating the feed-forward controller, and the driver changed the gear according to the suggestion by the haptic cue accelerator.

Figure 9.33 and Figure 9.34 show the results of fuel consumption. As shown in Figure 9.33, the mass flow rates of air-fuel charge into the cylinder were evaluated for each gear change. As clearly observed from the results, the case of haptic cue consumes much smaller air-fuel for all gear changes. The consumed amount of the air-fuel mixture can be calculated by integrating the mass flow rate over the driving time. The total amount of consumption decreased from 389.6 g to 333.9 g. If considering the fuel mixture ratio of 14.6, the amount of fuel consumption (only gasoline except air) can be calculated, as shown in Figure 9.34. From the results, the vehicle adopting the proposed haptic cue accelerator can save gasoline about 3.6 g for 48 s. Thus, it is expected that the proposed system can be quite satisfactory in real field conditions.

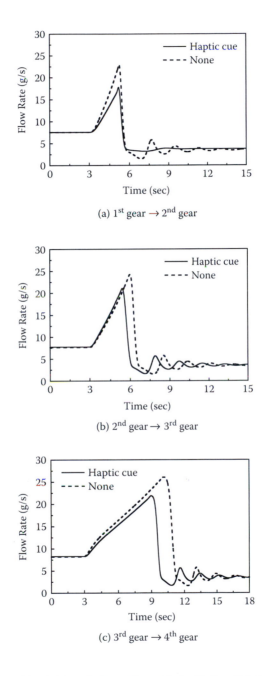

FIGURE 9.33
Air-fuel charge. (From Han, Y.M. et al., *Proceedings of the Institution of Mechanical Engineers : Part D - Journal of Automobile Engineering,* 225, 3, 2011. With permission.)

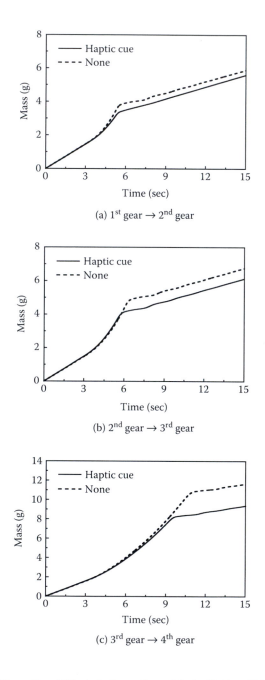

(a) 1st gear → 2nd gear

(b) 2nd gear → 3rd gear

(c) 3rd gear → 4th gear

FIGURE 9.34
Fuel consumption. (From Han, Y.M. et al., *Proceedings of the Institution of Mechanical Engineers : Part D - Journal of Automobile Engineering*, 225, 3, 2011. With permission.)

9.4 Some Final Thoughts

In this chapter, two fascinating haptic devices utilizing magnetorheological (MR) fluid were introduced for vehicle applications. One is the multi-functional haptic control knob for vehicular instrument control. This allows the driver and front-seat passenger to control such amenities as the climate (air conditioner and heater), the audio system (radio and CD player), the navigation system, communication system, and so on. The other is the haptic cue accelerator for vehicle manual transmission system. This concept of haptic cue accelerator can provide optimal gear shifting, which can reduce fuel consumption in economic driving or increase driving power in sport driving according to driver's preference, for manual transmission vehicles.

In both cases, the geometric dimension was optimally obtained to maximize a relative control torque by using the ANSYS parametric design language. The manufactured devices were interacted with a virtual vehicle that was constructed by dSPACE and MATLAB Simulink®. The virtual vehicle for the multi-functional haptic control knob emulated various events of the in-vehicle functions including menu shifting, temperature adjusting, and window opening. The virtual vehicle for the haptic cue accelerator emulated four-cylinder four-stroke engine, manual transmission, and small-sized car body. A simple feed-forward controller was formulated based on the inverse model compensation algorithm. By demonstrating force-feedback control performance, it has been proved that the multi-functional haptic control knobs well emulated each different operation of in-vehicle functions. In addition, the haptic cue accelerator also successfully cued the driver to change gears by torque reflection through the accelerator pedal. It has been proved from the mass flow rates of air-fuel charge into the cylinder that energy can be saved over 10% by adopting the proposed haptic cue accelerator into the manual transmission vehicles.

Now, the haptic device, closely related with x-by-wire technology, becomes a significant component of the future electric or hybrid vehicles. This MR haptic technology in the automotive industry will give a new function for vehicle driving and driver's convenience, and replace the traditional mechanical control systems by using electromechanical actuators and human-machine interfaces.

References

[1] Rovers, A. F. 2002. Haptic feedback: a literature study on the present-day use of haptic feedback in medical robotics. *DCT Report nr.*, University of Technology.

[2] BMW, iDrive Controller, http://www.bmw.com/com/en/insights/technology/technology_guide/articels/idrive.html

[3] Bengtsson, P., Grane, C., and Isaksson, J. 2003. Haptic/graphic interface for in-vehicle comfort functions - a simulator study and an experimental study. *Proceedings of the 2nd IEEE International Workshop on Haptic, Audio and Visual Environments and Their Applications*, 25–29.

[4] Aoki, J. and Murakami, T. 2008. A method of road condition estimation and feedback utilizing haptic pedal. *AMC'08 The 10th International Workshop on Advanced Motion Control*, Centro Santa Chiara, Trento, Italy, 777–782.

[5] Kobayashi, Y., Kimura, T., Yamamura, T., Naito, G., and Nishida, Y. 2006. Development of a prototype driver support system with accelerator pedal reaction force control and driving and braking force control. *SAE*, 2006-01-0572.

[6] Kenaley, G. L. and Cutkosky, M. R. 1989. Electrorheological fluid-based robotic fingers with tactile sensing. *Proc. 1989 IEEE Int. Conf. on Robotics and Automation*, Scottsdale, AZ, 132–136.

[7] Lee, H. S. and Choi, S. B. 2001. Control and response characteristics of a magneto-rheological fluid damper for passenger vehicles. *Journal of Intelligent Material Systems and Structures* 11: 80–87.

[8] Hong, S. R., Choi, S. B., Jung, W. J., and Jeong, W. B. 2003. Vibration isolation of structural systems using squeeze-mode ER mounts. *Journal of Intelligent Material Systems and Structures* 13: 421–424.

[9] Choi, S. B., Hong, S. R., Cheong, C. C., and Park, Y. K. 1999. Comparison of field-controlled characteristics between ER and MR clutches. *Journal of Intelligent Material Systems and Structures* 10: 615–619.

[10] Neelakantan, V. A. and Washington, G. N. 2005. Modeling and reduction of centrifuging in magnetorheological (MR) transmission clutches for automotive applications. *Journal of Intelligent Material Systems and Structures* 16: 703–712.

[11] Scilingo, E. P., Sgambelluri, N., Rossi, D. D., and Bicchi, A. 2003. Haptic displays based on magnetorheological fluids: design, realization and psychophysical validation. *Proceedings of the 11th Symposium on Haptic Interfaces for Virtual Environment and Teleoperator Systems*, Los Angeles, CA, 10–15.

[12] Li, W. H. 2004. Magnetorheological fluids based haptic device. *Emerald Sensor Review* 24: 68–73.

[13] Han, Y. M., Kim, C. J., and Choi, S. B. 2009. A magnetorheological fluids-based multifunctional haptic device for vehicular instrument controls. *Smart Materials and Structures* 18: 1–11.

[14] Han, Y. M., Noh, K. W., Lee, Y. S. and Choi, S. B. 2010. Magnetorheological haptic cue accelerator for manual transmission vehicle. *Smart Materials and Structures* 19: 1–10.

[15] Han, Y. M. and Choi, S.B. 2011. Performance evaluation of a MR haptic cue accelerator on the basis of an engine-transmission model. *Proceedings of the Institution of Mechanical Engineers : Part D - Journal of Automobile Engineering* 225: 281–293.

[16] Haj-Fraj, A. and Pfeiffer, F. 2001. Optimal control of gear shift operations in automatic transmissions. *Journal of the Franklin Institute-Engineering and Applied Mathematics* 338: 371–390.

[17] Sakaguchi, S., Kimura, E., and Yamamoto, K. 1999. Development of an engine-CVT integrated control system. *SAE*, 1999-01-0754.

[18] Kim, D. K., Peng, H., Bai, S., and Maguire, J. M. 2007. Control of integrated powertrain with electronic throttle and automatic transmission. *IEEE Transactions on Control Systems Technology* 15: 474–482.

[19] Nguyen, Q. H., Han, Y. M., Choi, S. B., and Wereley, N. M. 2007. Geometry optimization of MR valves constrained in a specific volume using the finite element method. *Smart Materials and Structures* 16: 2242–2252.

[20] Crossley, P. R. and Cook, J. A. 1991. A nonlinear engine model for drive train system development. *IEE International Conference Control* 91: 921–925.

[21] Beydoun, A., Wang, L. Y., Sun, J., and Sivashankar, S. 1998. Hybrid control of automotive powertrain systems: a case study. *Hybrid Systems: Computation and Control* 1386: 33–48.

[22] Noh, K. W., Han, Y. M., and Choi, S. B. 2009. Design and control of haptic cue device for accelerator pedal using MR fluids. Master Thesis, Inha University.

Index